# MOBILE AD HOC NETWORKS

# Mobile Ad Hoc Networks

## Energy-Efficient Real-Time Data Communications

*by*

BULENT TAVLI
*University of Rochester, NY, U.S.A.*

and

WENDI HEINZELMAN
*University of Rochester, NY, U.S.A.*

A C.I.P. Catalogue record for this book is available from the Library of Congress.

ISBN-10 1-4020-4632-4 (HB)
ISBN-13 978-1-4020-4632-2 (HB)
ISBN-10 1-4020-4633-2 ( e-book)
ISBN-13 978-1-4020-4633-9 (e-book)

---

Published by Springer,
P.O. Box 17, 3300 AA Dordrecht, The Netherlands.

*www.springer.com*

*Printed on acid-free paper*

Printed in the Netherlands.

*To our families*

# Contents

# List of Figures

# List of Tables

# Preface

The challenge in the design of a protocol architecture for Mobile Ad Hoc Networks (MANETs) is to efficiently convey information using an unreliable physical channel within a dynamic connected set of mobile, limited-range, limited-energy radios without the support of any infrastructure. Since a MANET is a dynamic, distributed entity, the optimal control of such a system should also be dynamic and adaptive. The global optimal solution for the coordination of a dynamic distributed network (*i.e.*, centralized control) can be achieved by continuously monitoring the global network status, which is not realizable, or at least not scalable, due to the overhead required to obtain such information. Although distributed coordination is realizable and practical, due to the lack of reliable coordination, its performance becomes unstable as the network load increases, and it cannot avoid the waste of valuable resources such as bandwidth and energy. We believe that a protocol architecture for MANETs that coordinates channel access through an explicit collective decision process based on available local information will outperform completely distributed approaches under a wide range of operating conditions in terms of throughput and energy efficiency without sacrificing the practicality and scalability of the architecture, unlike centralized approaches.

This research monograph presents the *Time Reservation using Adaptive Control for Energy efficiency (TRACE)* family of protocol architectures that achieve such coordinated channel access in a distributed manner for real-time data communications in MANETs. The TRACE protocols include Single-Hop TRACE (SH-TRACE), a time-frame based Medium Access Control (MAC) protocol for single-hop networks; Multi-Hop TRACE (MH-TRACE), which adds coordination in a multi-hop environment to the SH-TRACE protocol; Network-wide Broadcasting through TRACE (NB-TRACE), which incorporates network-wide broadcasting into the TRACE framework, and Multicasting through TRACE (MC-TRACE), which extends the TRACE framework to multicasting. Extensive simulations and theoretical analysis have shown that the

TRACE protocols outperform non-coordinated network protocols in terms of energy efficiency without sacrificing the spatial reuse efficiency and the quality of service requirements of the application layer.

This monograph originated from a body of work [1] that was performed by Dr. Bulent Tavli at the University of Rochester during his Ph.D. program with Prof. Wendi Heinzelman. During this effort, it became clear to the authors that energy efficiency and Quality of Service (QoS) support are already significant issues in current wireless network technologies and that the importance of energy efficiency and QoS support will only increase in future wireless network technologies.

## Acknowledgments

This book was made possible in part by the Center for Electronic Imaging Systems (CEIS), a New York State Office of Science, Technology, and Academic Research (NYSTAR) designated center for advanced technology, and in part by the Harris Corporation, RF Communications Division. We would like to thank Tolga Numanoglu for his contributions to Chapter 8.

BULENT TAVLI AND WENDI HEINZELMAN

# Chapter 1

# INTRODUCTION

The era of wireless communications began with the first successful demonstration of wireless information transmission by Nikola Tesla in 1893 [2]. Although wireless communication techniques have been in use since then, it was not until the last decade of the twentieth century that wireless communication (*e.g.*, cell phones) become ubiquitous. Compared with conventional wired networks, the advantages of the cellular system include a reduction in infrastructure requirements and support for mobile communications. Encouraged by the success of the cellular revolution, the goal of communication researchers has been to achieve communications without relying on a fixed infrastructure. The goal is to create a network that has similar performance to a cellular system, even to conventional wired networks, without requiring any infrastructure support. This is the basic philosophy behind Mobile Ad Hoc Networks (MANETs). Although the military has been using multi-hop ad hoc networks for a long time, there are not yet many commercial applications for MANETs. However, the ultimate target, which is zero infrastructure mobile networking, is so enticing that government, industry, and academia have focused a great deal of time and effort to make this vision a reality.

The challenge in the design of protocol architectures for MANETs is to efficiently convey information using an unreliable physical channel within a highly dynamic set of mobile, limited-range, limited-energy, half-duplex radios without the support of any infrastructure. An efficient network protocol should jointly optimize the throughput, delay, and energy dissipation of the network without sacrificing fairness, robustness, and Quality of Service (QoS). However, the aforementioned set of design goals is a collection of contradicting metrics, suggesting that tradeoffs are required in the design of protocol architectures. Since a mobile ad hoc network is a highly dynamic, distributed entity the optimal control/coordination of such a system should also be highly dynamic and adaptive. The global optimal solution for the coordination of a dynamic distributed network (*i.e.*, centralized control) can be achieved by continuously monitoring the global network status, but this approach is not scalable due to the overhead required to obtain such information. Although distributed coordination is realizable and practical, due to the lack of reliable coordination, it is highly unlikely that distributed control could overcome instability and the underutilization and waste of valuable resources such as bandwidth and energy. Furthermore, without explicit coordination, which necessitates local coordinators, a network protocol cannot quickly adapt to dynamically changing conditions, such as spatial and/or temporal variations in traffic, node density, and mobility.

Our design philosophy is that a protocol architecture for MANETs that coordinates channel access through an explicit collective decision process based on available local information will outperform non-coordinated approaches under a wide range of operating conditions in terms of throughput and energy efficiency without sacrificing the practicality and scalability of the architecture, unlike the centralized approaches.

## 1.1 Characteristics of MANETs

A MANET is an autonomous system of mobile nodes with routing capabilities connected by wireless links, the union of which forms a communication network. A MANET can either be a standalone entity or it can be an extension of a wired network. There are many application areas of MANETs, such as:

- Military tactical operations - for fast and possibly short term establishment of military communications for troop deployments in hostile and/or unknown environments.
- Search and rescue missions - for communication in areas with little or no wireless infrastructure support.
- Disaster relief operations - for communication in environments where the existing infrastructure is destroyed or left inoperable.
- Law enforcement - for secure and fast communication during law enforcement operations.
- Commercial use - for enabling communications in exhibitions, conferences, and large gatherings.

The perception that a wireless ad hoc network is equivalent to a conventional tethered network except that the cables are replaced with antennas is a common misconception. Wireless ad hoc networks have unique characteristics that necessitate special solutions. Some of these differences are: (*i*) unreliable, half-duplex physical channel, (*ii*) dynamic topology changes, (*iii*) limited bandwidth, and (*iv*) limited energy resources. Thus, the wealth of knowledge in the area of conventional networking cannot be applied directly to wireless ad hoc networks.

When compared to an ordinary cable interface, wireless physical channels are very noisy and their bit error rates are much higher; thus packet losses are not uncommon, and, network protocols cannot be designed on the assumption of perfect transmissions/receptions. For example, a protocol should be equipped with mechanisms to recover from frequent packet losses. Note that the corrupted packets are not only the data packets but also the control packets that network protocols rely on to coordinate network operation.

Wireless radios are half-duplex, which means that they cannot receive while transmitting. Thus, collision detection by a transmitting node is impossible, which is the main reason that the Ethernet protocol cannot be used in wireless communications. The reason for this behavior is that the dynamic range in wireless communication is too high to enable a transmitting radio to detect any other transmissions; the receiver of a transmitting radio is already jammed by the interference created by its own transmission.

Node mobility, natural (*e.g.*, trees, hills) or man made (*e.g.*, buildings, walls) barriers in or near the propagation paths, and environmental (*e.g.*, rain, snow) or electronic (*e.g.*, microwave ovens, radio stations, military jamming) interference affecting the propagation characteristics all manifest themselves as dynamic topology changes, which directly or indirectly change the connectivity pattern of the network. Unlike in wired networks, where network topologies do not change frequently, even without node mobility wireless networks are highly dynamic. Therefore, a wireless network protocol has an additional burden when compared to a wired network protocol, which is mobility management and topology maintenance. Both of these are necessary to keep the wireless network as an organized distributed entity, which otherwise would not be useful for reliably conveying information.

Unlimited bandwidth is not available either in wired or in wireless networks. However, the available bandwidth for wireless networks is much less than that of wired networks. Furthermore, the protocol overhead in wireless networks is much higher in order to compensate for the unreliable channel and to maintain the network topology, which is required for routing.

The assumption of mobility, especially the mobility of pedestrians, suggests that the radios be lightweight, and thus they cannot have a large energy supply. A limited energy supply necessitates avoidance of energy waste. Energy efficiency of a network can be achieved by the collective collaboration of the physical layer (*i.e.*, hardware), medium access control layer, network layer, and upper layers. In other words, a cross-layer design is needed to achieve optimal energy efficiency of a protocol architecture.

## 1.2 Importance of QoS and Energy Efficiency in MANETs

Having summarized the unique characteristics of MANETs, we will focus on the specific area of this book - *energy-efficient real-time group communications in MANETs*. Real-time voice communication is commonly used in many MANET scenarios that include groups of people with no available infrastructure support. However, both the efficiency and the versatility of these applications suffer seriously due to the lack of an underlying network protocol designed specifically for energy-efficient real-time group communications.

There is a considerable accumulation of research on all major components of energy-efficient real-time group communications in MANETs: (*i*) energy-efficient protocol design, (*ii*) real-time voice communications, and (*iii*) group communications (broadcasting and multicasting) in ad hoc networks. However, a multi-objective protocol architecture design for (*i*) minimizing energy dissipation, (*ii*) providing QoS for voice packets, and (*iii*) enabling efficient multi-hop broadcasting and multicasting has not been thoroughly investigated in the literature.

Providing QoS for multimedia traffic (*e.g.*, voice) has been a design objective for many wireless network protocols [3–9]. Most of these protocols are designed either for single-hop networks or have QoS provisions in single-hop configurations, where a certain level of infrastructure is required. There are also a few protocol architectures [10, 11] that provide QoS in multi-hop networks. However, providing QoS in broadcasting or multicasting has found little attention. The main reason for this lack of attention is that multi-hop broadcasting or multicasting has been considered only as a tool for unicasting [12] (*i.e.*, route discovery, topology exchange, *etc.*). However, due to advances in technology and the understanding and maturity of multi-hop ad hoc networks, applications that require voice broadcasting and multicasting are becoming important, and new protocols are needed to support this service.

Broadcasting and multicasting for data communications have also been investigated extensively in the literature [11–20]. However, broadcasting and multicasting voice packets have some unique constraints, such as QoS, which necessitates special treatment. For the same reasons described previously, voice broadcasting and multicasting in MANETs have not been investigated extensively in the past.

Popular network architectures, such as IEEE 802.11 and Bluetooth, include mechanisms to save energy [3, 8]. However, these provisions are not specifically for voice communications, and they often contradict the QoS requirements of the application (*i.e.*, delay/energy dissipation tradeoff). Some protocol architectures, such as IEEE 802.15.3 [9], include mechanisms for saving energy without violating the QoS of multimedia applications. However, these protocols are only designed to operate efficiently in single-hop networks.

There are several protocol architectures that modify existing ad hoc network protocols for energy efficiency [21, 22]. However, these protocols are either designed for specific applications other than voice or their energy savings are very low.

In light of the preceding discussion, it is clear that energy-efficient real-time group communications in MANETs is an important design problem that has not been investigated sufficiently in the past. In this book, we present our design, analysis, and simulation of the Time Reservation using Adaptive Control for Energy efficiency (TRACE) family of protocol architectures for energy-efficient real-time data communications in MANETs.

## 1.3 Scope and Novelty of the Book

We have developed the TRACE family of protocol architectures for energy-efficient real-time voice communications in wireless ad hoc networks. The common features of these protocol architectures are: (*i*) coordinated channel access through clustering and scheduling for dynamic switching between the sleep/active modes for energy efficiency and stability, (*ii*) cyclic time-frame based channel access for QoS support, (*iii*) information summarization prior to actual data transmission for energy efficiency, (*iv*) distributed system design for scalability, and (*v*) reliability and fault tolerance for robustness. We conducted extensive mathematical and simulation analysis of these protocols under varying network conditions and parameters with several application scenarios. Furthermore, we compared the TRACE protocols with many existing protocols through careful quantitative and qualitative analysis. We also investigated the broadcast capacity of wireless networks and derived an asymptotic upper bound. Contributions of these research efforts to the state-of-the-art are itemized below.

- A cyclic time-frame based MAC protocol (SH-TRACE) designed primarily for energy-efficient, reliable, real-time voice packet broadcasting in a peer-to-peer, single-hop infrastructureless radio network is presented.
- A MAC protocol that combines advantageous features of fully centralized and fully distributed networks for energy-efficient real-time packet broadcasting in a multi-hop radio network (MH-TRACE) is designed.
- Coordinated channel access, managed by local coordinators, greatly reduces data packet collisions in multi-hop networks, especially in high node density and/or high data rate networks. Furthermore, data packet collisions are completely eliminated in fully connected networks through explicit coordination of the channel access by dynamically selected coordinators.
- Transparent clustering completely removes the hard boundaries in a multi-hop network commonly encountered in clustered ad hoc networks.

- Significant energy savings are achieved by using information summarization prior to data transmission, eliminating idle listening, collision reception, and unnecessary carrier sensing.

- Receiver-based listening cluster creation is shown to be a highly energy-efficient method for data discrimination in group communications.

- Cyclic time-frame based automatic channel access, which has been shown to be an effective way of providing QoS in single-hop cellular systems, has been efficiently extended to multi-hop clustered ad hoc networks.

- A novel, simple, and distributed framework for clustering and inter-cluster interference avoidance is created.

- Energy efficiency and resilience against channel errors for coordinated and non-coordinated MAC protocols are investigated through simulations and analysis. We have shown that it is possible to achieve better system performance with coordinated MAC protocols even in lossy channels, provided that the BER level is not extremely high.

- A detailed performance evaluation of MH-TRACE and other MAC protocols when they are used for network-wide voice broadcasting through flooding is performed through extensive simulations. Furthermore, it is shown that MH-TRACE energy efficiency is superior to other MAC protocols in network-wide voice broadcasting through flooding. In addition, it is shown that the dominant energy dissipation term in this application for Carrier Sense Multiple Access (CSMA)-based architectures is carrier sensing energy dissipation, and transmit energy dissipation is just a minor component of the total energy dissipation.

- Energy- and spatial reuse-efficient QoS supporting network-wide broadcasting and multicasting architectures (NB-TRACE and MC-TRACE) based on MH-TRACE are designed and analyzed. These are the first examples of network-wide broadcasting/multicasting architectures that reduce the total energy dissipation rather than the transmit energy dissipation only.

- Information summarization is shown to be a very effective means of avoiding energy dissipation on redundant data retransmissions, which are inherently difficult to eliminate in broadcasting.

- Automatic renewal of channel access, primarily used in fully-connected single-hop networks, is reengineered as a bandwidth reservation and broadcast/multicast tree creation and maintenance mechanism, which results in virtually zero jitter and high spatial reuse efficiency.

- A multi-stage contention algorithm that results in a maximal number of successful contentions in minimum time for S-ALOHA type contention systems is presented.

- An asymptotic upper bound for the broadcast capacity of wireless ad hoc networks is established. Unlike unicasting, where per node capacity in an $n$-node network is shown to be bounded by $O(1/\sqrt{n})$, in broadcasting the per node broadcast capacity is shown to be bounded by $O(1/n)$.

## 1.4 High Level Overview of the Book

The first part of this book (Chapters 2 through 5) provides background information on mobile ad hoc networking, in general, and energy efficiency and QoS-support, in particular. Chapter 2 presents the fundamental concepts of networking with a strong emphasis on wireless ad hoc networks. Most of our research is focused on the MAC and routing layers, hence, we devote Chapter 3 and Chapter 4 to descibing different MAC and routing layer protocols, respectively. This helps to establish a sufficient level of understanding of the problems and solutions for general mobile ad hoc networks, which will be helpful for the following chapters. Having equipped ourselves with the necessary information on wireless adhoc networking, Chapter 5 is written as a bridge to the later chapters of this book. Chapter 5 describes the energy dissipation characteristics of wireless networks and presents a literature review on energy saving methodologies in wireless networks. Chapter 5 also introduces the QoS requirements for voice communications and QoS supporting protocol architectures. The aforementioned background information is provided to create a concise foundation for the later material of this book. However, the reader can use the references provided in these chapters to acquire a broader understanding of wireless communications and networking.

The second part of this book (Chapters 6 through 12) presents our research on the design and analysis of the TRACE family of protocol architectures for energy-efficient voice communications in mobile ad hoc networks (see Figure 1.1). Chapter 6 describes the Single-Hop TRACE (SH-TRACE) protocol architecture in detail, and presents the simulation results, theoretical analysis, and comparisons with well known protocol architectures. Principles, extensive simulations, and theoretical analysis of the Multi-Hop TRACE (MH-TRACE) protocol architecture are presented in Chapter 7. We analyze the effects of channel noise on the performance of the MH-TRACE protocol architecture in Chapter 8. A comparison of MH-TRACE and several other MAC protocols for real-time data broadcasting through flooding is presented in Chapter 9. The Network-wide Broadcasting through TRACE (NB-TRACE) protocol architecture design principles, motivations, and limitations are presented in Chapter 10. The MultiCasting through TRACE (MC-TRACE) protocol architecture is

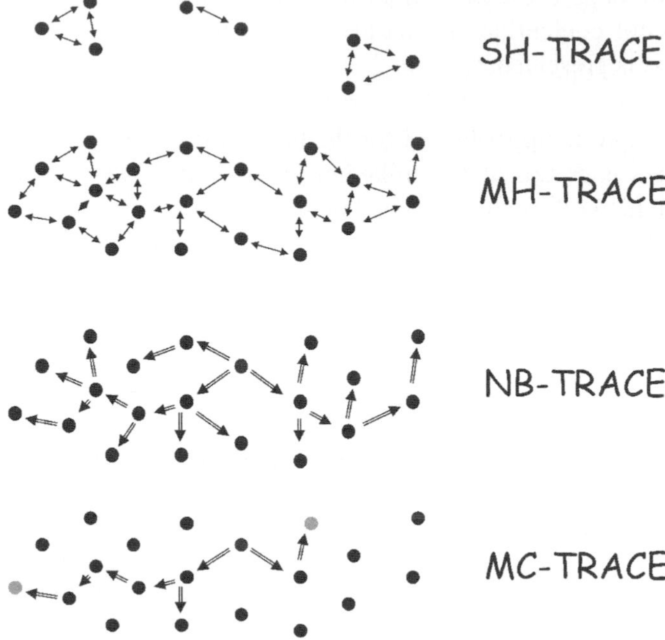


*Figure 1.1.* TRACE family of protocol architectures.

presented and analyzed in Chapter 11. Finally, we present general conclusions and di scuss future research directions in Chapter 12.

An algorithm for optimizing the contention stage of the SH-TRACE protocol is presented in Appendix A. The effects of constraining inter-cluster separation on the performance of MH-TRACE are presented in Appendix B. We present an asymptotic upper bound on the broadcast capacity of wireless ad hoc networks in Appendix C. A glossary of the terms used in this book is presented in Appendix D.

# Chapter 2

# MANET FUNDAMENTALS

In this chapter we present an introduction to Mobile Ad Hoc Networks (MANETs). As the emphasis of this book is on the networking component of MANETs, we start with the fundamentals of networking.

A communication protocol is a predetermined set of rules or conventions that enables efficient communications within a computer network by providing a well-defined structure for information exchange between hosts. It has both syntactic (*e.g.*, packet types and formats) and semantic (*e.g.*, handling a corrupted packet) components. In the following section we will introduce the performance metrics for designing efficient MANET protocols.

## 2.1 Performance Metrics

MANET protocols are the key elements in determining many features of a wireless network, such as throughput, Quality of Service (QoS), energy dissipation, fairness, stability, and robustness [23] (see Figure 2.1). The following is a brief discussion of these metrics.

- Throughput - In the context of communication networks, throughput is defined as the fraction of the raw bandwidth used exclusively for data transmission. It is not possible to use 100% of the bandwidth for data transmissions due to the unavoidable bandwidth used for overhead (*e.g.*, packet headers, control packets, guard bands). The objective of MANET protocols is to keep the bandwidth used for overhead as low as possible (high throughput) without sacrificing the other objectives.

- QoS - Low delay, high packet delivery ratio, and guaranteed bandwidth are some of the metrics that can define QoS, which is an application-dependent concept. For example, QoS for voice packets consists of three components: (*i*) high packet delivery ratio, (*ii*) low delay, and (*iii*) low jitter. Since

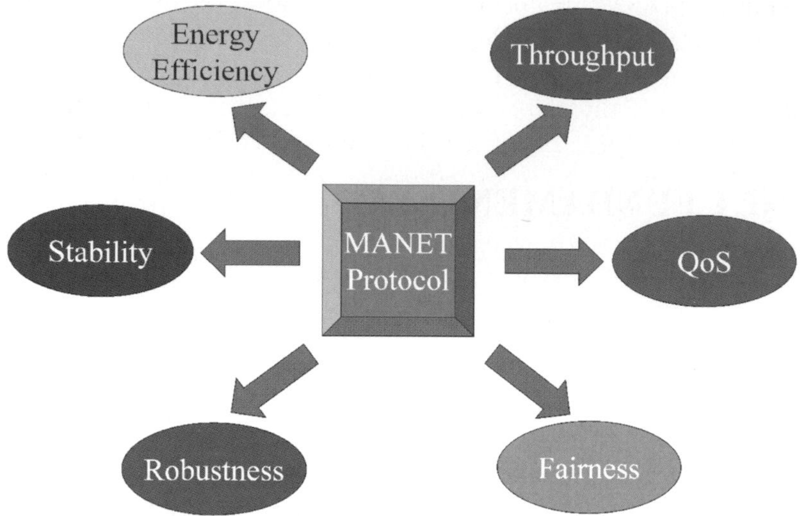


*Figure 2.1.* MANET protocol performance metrics.

voice packets are created periodically, MANET protocols should be able to grant periodic channel access for the voice sources without violating the maximum allowable delay for the voice packets, after which the voice packets are dropped.

- Energy dissipation - Energy efficiency is crucial for lightweight battery-operated radios to avoid consuming their limited energy resources. Idle listening is an important energy dissipation term, which can be avoided by switching the radio to a low energy sleep mode. Since in sleep mode a radio cannot receive or transmit, MANET protocols should have mechanisms to seamlessly put the radio into the sleep mode and take it back to the active mode without violating the efficient operation of the network.

- Fairness - Maximization of throughput can be achieved by letting a single node transmit indefinitely, which results in unfairness against the rest of the nodes in the network. Fairness can be achieved by partitioning the network resources (*i.e.*, bandwidth) in a balanced fashion among the nodes trying to obtain channel access. For example, in a network with 1 Mbps bandwidth and nodes A and B with bandwidth requirements of 0.4 Mbps and 0.6 Mbps, respectively, the channel allocations should be 0.4 Mbps for node A and 0.6 Mbps for node B. Thus fairness is more than simple division of the bandwidth into equal shares. MANET protocols are responsible for granting channel access fairly among the users in a dynamic fashion.

- Stability - MANET protocols control a dynamic system, thus their performance can become unstable, like many dynamic systems, if certain conditions are not met. It is a well-known fact that many protocols such as ALOHA and IEEE 802.11 can become unstable if the demand for channel access is higher than some threshold value. Unless otherwise noted, throughout this book, "IEEE 802.11" is used for IEEE 802.11 in ad hoc mode. Stable MANET protocols should be able to avoid instability.

- Robustness - It is not uncommon to loose packets in wireless communications, and some of the lost packets are the control packets used by the MANET protocols themselves. Some MANET protocols are based on centralized control through coordinator nodes. It is possible that these nodes can be left inoperable (*e.g.*, their batteries run out). Thus, robust MANET protocols should be designed to continue their normal operation without becoming unstable under packet losses or node failures.

## 2.2 The Layered Communication Network

Most of the popular network protocols are created with a modular design methodology, where the modules of a synthesized protocol are arranged in a vertical stack. The protocol stack is a generic model of the organization of a layered communication system. There are several reference models for describing the layers of a communication network, such as the Open System Interconnection (OSI) reference model [24] and the Transmission Control Protocol/ Internet Protocol (TCP/IP) reference model [25]. The objective for organizing the network interface into layers is simple and clear: management of a single complex module is not easy as a general design rule in the broad field of technology. Instead, a system created from well-integrated but separable blocks is easier to design, manage and maintain.

To emphasize the functionality of various layers of a generic communication protocol, we will focus on the layered protocol stack described in [26], which is basically the TCP/IP reference model and is shown in Figure 2.2.

### 2.2.1 The Channel

The channel is the medium that is used to convey the information. For example, the channel could be coaxial or fiber optic cables in wired networks, electromagnetic waves in wireless networks or satellite systems, or the combination of different types of medium.

In a wireless channel, the electromagnetic wave power can be modeled as falling off as a power law function of the distance between the transmitter and receiver. In addition, if there is no direct, line-of-sight path between the transmitter and the receiver, the electromagnetic wave will bounce off objects in the environment and arrive at the receiver from different paths at different

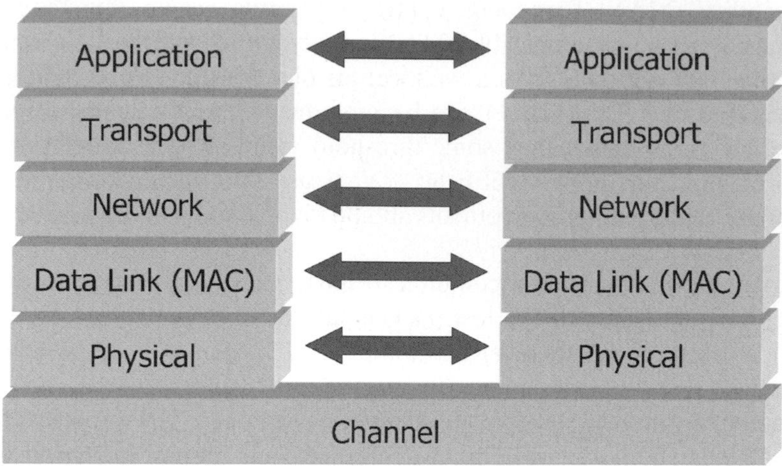


*Figure 2.2.* TCP/IP reference model.

times. This causes multipath fading, which again can be roughly modeled as a power law function of the distance between the transmitter and receiver. No matter which model is used (direct line-of-sight or multipath fading), the received power decreases as the distance between the transmitter and receiver increases [27].

For the experiments described in this book, both the free space model and the multipath fading model were used, depending on the distance between the transmitter and receiver, as defined by the channel propagation model in `ns` [28, 27]. If the distance between the transmitter and receiver is less than a certain cross-over distance ($d_{crossover}$), the free space model is used ($d^2$ attenuation), and if the distance is greater than $d_{crossover}$, the two-ray ground propagation model (a multipath fading model) is used ($d^4$ attenuation). The cross-over point is defined as follows:

$$d_{crossover} = \frac{4\pi\sqrt{L}h_r h_t}{\lambda} \tag{2.1}$$

where

$L \geq 1$ is the system loss factor not related to propagation,

$h_r$ is the height of the receiving antenna above the ground,

$h_t$ is the height of the transmitting antenna above the ground, and

$\lambda$ is the wavelength of the carrier signal.

If the distance is less than $d_{crossover}$, the transmit power is attenuated according to the free space equation as follows:

$$\wp_r(d) = \frac{\wp_0 G_t G_r \lambda^2}{(4\pi d)^2 L} \tag{2.2}$$

where

$\wp_r(d)$ is the receive power given a transmitter-receiver separation of $d$,

$\wp_0$ is the transmit power,

$G_t$ is the gain of the transmitting antenna,

$G_r$ is the gain of the receiving antenna,

$\lambda$ is the wavelength of the carrier signal,

$d$ is the distance between the transmitter and the receiver, and

$L \geq 1$ is the system loss factor not related to propagation.

This equation models the attenuation when the transmitter and receiver have direct, line-of-sight communication, which will only occur if the transmitter and receiver are close to each other (*i.e.*, $d < d_{crossover}$). Note that the transmit power, $\wp_0$, is the net power at the antenna. The actual power drawn from the battery is larger than $\wp_0$ with the rest of the power used by the electronic circuitry before reaching the antenna. If the distance between the transmitter and the receiver is greater than $d_{crossover}$, the transmit power is attenuated according to the two-ray ground propagation equation as follows:

$$\wp_r(d) = \frac{\wp_0 G_t G_r h_t^2 h_r^2}{d^4} \tag{2.3}$$

where

$\wp_r(d)$ is the receive power given a transmitter-receiver separation of $d$,

$\wp_0$ is the transmit power,

$G_t$ is the gain of the transmitting antenna,

$G_r$ is the gain of the receiving antenna,

$h_r$ is the height of the receiving antenna above the ground,

$h_t$ is the height of the transmitting antenna above the ground, and

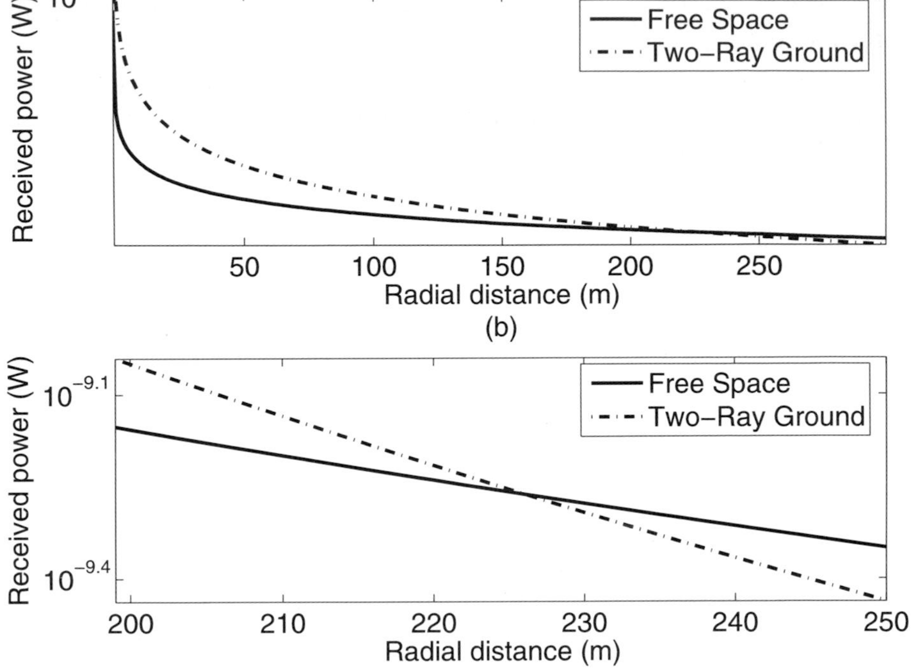


*Figure 2.3.* Free Space and Two-Ray Ground propagation models.

$d$ is the distance between the transmitter and the receiver.

In this case, the received signal comes from both the direct path and a ground-reflection path [27]. Due to destructive interference when there is more than one path through which the signal arrives, the signal is attenuated as $d^4$.

In the experiments described in this book, an omnidirectional antenna was used with the following parameters: $G_t = G_r = 1$, $h_t = h_r = 1.5$ m, no system loss ($L = 1$), 2.4 GHz radios, and $\lambda = \frac{3\times10^8}{2.4\times10^9} = 0.125$ m. Using these values, $d_{crossover} = 226.2$ m and Equations 2.2 and 2.3 simplify to (see Figure 2.3):

$$\wp_r = \begin{cases} 0.99 \times 10^{-4} \frac{\wp_0}{d^2} & : \; d < 226.2 \text{ m} \\ 5.06 \frac{\wp_0}{d^4} & : \; d \geq 226.2 \text{ m} \end{cases} \tag{2.4}$$

### 2.2.2 Physical Layer

The physical layer is the modem hardware in simple terms. For example, the antenna and the transmitter/receiver electronics are parts of the physical layer in a wireless node. Modulation and coding are the main functions of the physical layer. Furthermore, many other auxiliary services (*e.g.*, clear channel assessment and receive power signal strength monitoring) are also provided for the upper layers by the physical layer.

The bitstream coming down from the upper layers cannot be conveyed efficiently through the channel directly, thus, the bits or chunks of bits need to be transformed into a more efficient representation that can be optimally conveyed through the channel. Effective radiation of electromagnetic waves requires antenna dimensions comparable with the wavelength. For example, an antenna for a 3 kHz carrier would be 100 km long and an antenna for 3 GHz carrier is 10 cm long.

The operation of mapping the raw digital data (*i.e.*, zeros and ones) onto carrier signals is known as modulation and the inverse of this operation is known as demodulation. There are many modulation schemes used in wireless communications. We will present a short summary of these modulation schemes.

The simplest form of carrier modulation is Amplitude Shift Keying (ASK). In ASK, a binary one is represented by the presence of a non-zero constant amplitude cosine function and a zero is represented by the absence of the carrier (see Figure 2.4). Thus the mapping in ASK can be represented by the following equation:

$$s_{ASK}(t) = \begin{cases} 0 & : \text{ binary 0} \\ Acos(2\pi f_0 t) & : \text{ binary 1} \end{cases} \tag{2.5}$$

where $s_{ASK}(t)$ is the signal in the analog domain. ASK modulation is used in some optical networks (in fiber or free-space) due to the practical and inexpensive transmitter and receiver implementations.

Representing a binary zero with the absence of the carrier creates problems in detection (*e.g.*, it is difficult to detect multiple contiguous binary zeros or to differentiate a binary zero from the no transmission state). These problems can be alleviated by assigning two distinct carriers to a binary zero and a binary one, which slightly increases the system complexity. This modulation scheme is known as Binary Frequency Shift Keying (BFSK), which is the simplest form of Frequency Shift Keying (FSK) modulation. The mapping in BFSK can be represented as follows:

$$s_{BFSK}(t) = \begin{cases} Acos(2\pi f_0 t) & : \text{ binary 0} \\ Acos(2\pi f_1 t) & : \text{ binary 1} \end{cases} \tag{2.6}$$

BFSK is less susceptible to additive white noise than ASK. Actually, in the alphabet of ASK there is only one symbol, which is the single carrier ($Acos(2\pi f_0 t)$). On the other hand, in the BFSK alphabet there are two symbols ($Acos(2\pi f_0 t)$ and $Acos(2\pi f_1 t)$), which makes BFSK more bandwidth efficient. It is possible to increase the bandwidth efficiency by increasing the alphabet size. However, this makes the reception of the symbols more error-prone due to the reduced separation among the symbols.

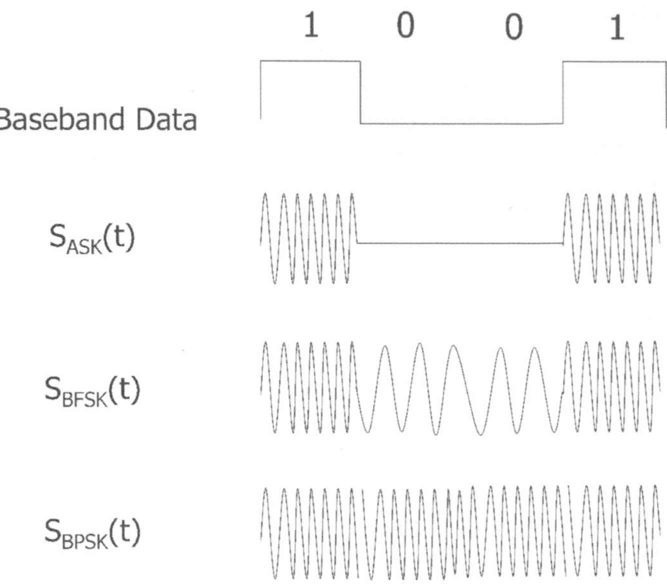


*Figure 2.4.* Illustration of ASK, BFSK, and BPSK.

In M-ary FSK (MFSK) there are more than two frequencies. Since, the symbol size is now larger than two, instead of assigning a single binary digit to each symbol, chunks of binary digits are assigned to each symbol. The mapping in MFSK can be represented as follows [2]:

$$s_{MFSK}(t) = \quad Acos(2\pi f_i t) \quad : \quad 1 \leqslant i \leqslant M \tag{2.7}$$

where $f_i = f_c + (2i - 1 - M) f_d$, $f_c =$ the carrier frequency, $f_d =$ the difference frequency, $M =$ the number of different symbols $= 2^L$, and $L =$ the number of bits per symbol. Table 2.1 presents the frequency assignment for each of the 8 possible 3-bit data combinations with $f_c = 250$ kHz, $f_d = 25$ kHz, and $M = 8$ ($L = 3$ bits).

It is possible to create an alphabet with a single carrier by mapping the binary digits into the phase of the carrier, which is called Phase Shift Keying (PSK). Thus, in PSK, the phase of the carrier signal is shifted to represent data. Binary PSK (BPSK) uses two phases to represent the two binary digits. The mapping in BPSK can be represented as follows [2]:

$$s_{BPSK}(t) = \begin{cases} Acos(2\pi f_0 t) \\ Acos(2\pi f_0 t + \pi) \end{cases} = \begin{cases} Acos(2\pi f_0 t) & : \text{ binary } 0 \\ -Acos(2\pi f_0 t) & : \text{ binary } 1 \end{cases} \tag{2.8}$$

*Table 2.1.* Frequency assignment in 8-FSK with $f_c = 250$ kHz, $f_d = 25$ kHz, and $M = 8$ ($L = 3$ bits)

| | | | |
|---|---|---|---|
| $f_1$ | = | 75 kHz | 000 |
| $f_2$ | = | 125 kHz | 001 |
| $f_3$ | = | 175 kHz | 010 |
| $f_4$ | = | 225 kHz | 011 |
| $f_5$ | = | 275 kHz | 100 |
| $f_6$ | = | 325 kHz | 101 |
| $f_7$ | = | 375 kHz | 110 |
| $f_8$ | = | 425 kHz | 111 |

In BPSK, the signal representing a binary zero ($Acos(2\pi f_0 t)$) is just the negative of the signal representing a binary one ($-Acos(2\pi f_0 t)$). The major drawback of BPSK is the rapid amplitude change between symbols due to phase discontinuity, which requires infinite bandwidth. However, BPSK demonstrates better performance than ASK and BFSK. BPSK can be expanded to an M-ary scheme, employing multiple phases and amplitudes as different states.

The bitstream from the upper layers is not coming directly from the data source (*i.e.*, it is not raw data). Instead, it goes through an encapsulation at each layer of the protocol stack, and it is fed to the physical layer in chunks of data. These data chunks are called packets. At each layer the packet coming from the upper layer is encapsulated and passed to the lower layer (see Figure 2.5). The Encapsulation process includes adding bits to the front (packet header) and possibly end of the packet. Furthermore, the bits of the payload (*i.e.*, all of the bits of the packet coming from the upper layer, which are treated as the payload for the lower layer) can be modified (*e.g.*, scrambling, adding redundant bits, *etc.*) in the process of encapsulation. In the reverse process (*i.e.*, packets passed to the upper layer) of decapsulation, all the effects of encapsulation are stripped from the packet and the payload is sent to the upper layer (*e.g.*, a network layer protocol removes the network protocol header from the packet and passes the payload up to the transport layer protocol).

### 2.2.3 Data Link Layer

The limited bandwidth of the wireless channel combined with radio propagation loss and the broadcast nature of radio transmission make communication over a wireless channel inherently unreliable. Link-layer protocols are used to add information bits to the data bits to protect them against channel errors. Forward error correction (FEC) protocols add controlled redundancy to the data, in order to enable reliable transmission of data over unreliable channels. Typical channel coding systems contain a source coder, to reduce (with the goal of

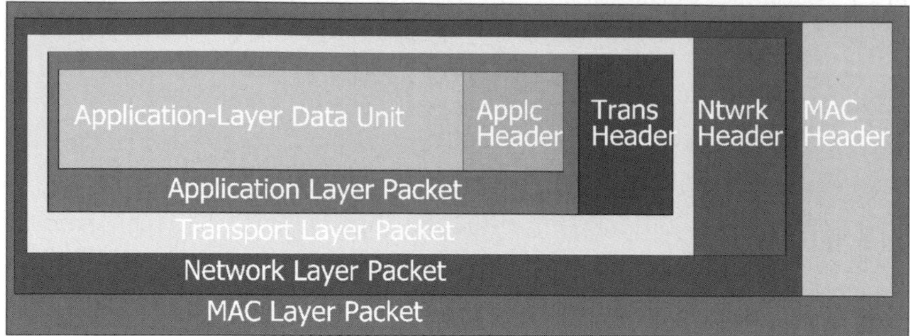


*Figure 2.5.* Successive encapsulation.

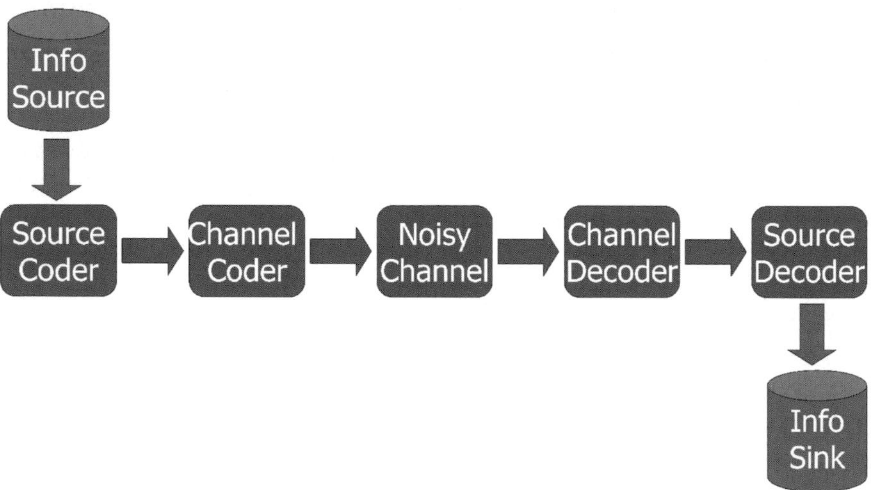


*Figure 2.6.* Block diagram of a data transmission system.

removing) redundancy from the data, followed by a channel coder, that adds controlled redundancy to the compressed data. The channel-coded data are sent over a channel, where noise is added to the stream. The channel decoder at the receiver produces an estimate of the source-coded stream, which is sent to the source decoder to extract the data to be given to the application. This system is depicted in Figure 2.6.

The two basic types of channel coders for FEC are block coders and convolutional coders. Block coders take a block of size $k$ and produce a coded block of size $n$ that depends only on the information in that block. This produces an $(n, k)$ block code, where there are $2^k$ possible input blocks and $2^n$ possible output blocks. Convolutional coders also take blocks of size $k$ as input and produce a coded block of size $n$. However, the output symbol is a function of not only the input block but of the last $m$ input blocks. This represents an $(n, k, m)$ convolutional code with memory order $m$ [29]. For both block and convolutional codes, the code rate is $R = \frac{k}{n}$.

Convolutional encoding of data is performed by convolving $k$ bits of the input with $n$ generator polynomials to produce a rate-$\frac{k}{n}$ code, where $k \leq n$. The added redundancy is used at the decoder to detect and correct a certain number of errors, using, for example, Viterbi decoding. Convolutional encoders are typically implemented using shift registers. If $k = 1$, the input stream can be continuously fed into the shift registers and the output can be continuously read at $\frac{n}{k}$ times the rate of the input.

Convolutional codes have the nice property that several different rate codes can be achieved using the same *mother code*, so a single hardware implementation can produce varying amounts of protection. A rate-$\frac{a}{b}$ code can be achieved by puncturing, or discarding, the output bits from a rate-$\frac{1}{n}$ code. For every $a$ input bits to the rate-$\frac{1}{n}$ coder, $(na - b)$ of these bits are discarded. The remaining $b$ bits are sent as the channel coded signal. Rate Compatible Punctured Convolutional (RCPC) [30] encoding is a special type of puncturing where higher rate codes are subsets of lower rate codes. Searches have been performed to determine the best generator and puncturing matrices [30].

### 2.2.4 Medium Access Control

The Medium Access Control (MAC) layer is normally considered as a sublayer of the data link layer in the protocol stack [31]. However, it deserves to be an independent layer due to its importance in wireless communications. The channel in wireless communications is a shared resource. Efficient utilization of this shared resource necessitates that there should be a set of rules and conventions among the nodes that share the same channel.

MAC protocols are used to create predefined ways for multiple users to share the channel. MAC protocols have been extensively studied in traditional areas of wireless voice and data communications [31]. There are two fundamentally different ways to share the wireless channel bandwidth among different nodes [32]: fixed-assignment channel-access methods (*e.g.*, Time Division Multiple Access (TDMA), Frequency Division Multiple Access (FDMA), Code Division Multiple Access (CDMA), and Space Division Multiple Access (SDMA)) and random access methods (*e.g.*, IEEE 802.11, Carrier Sense Multiple Access (CSMA), Multiple Access Collision Avoidance (MACA), and MACA for Wireless (MACAW)[33]). Fixed-assignment MAC protocols allocate each user a given amount of bandwidth, either slicing the spectrum in time (TDMA), frequency (FDMA), code (CDMA), or space (SDMA). These MAC protocols are widely used in modern cellular systems [27]. Since each node is allocated a unique part of the spectrum, there are no collisions among the data. However, fixed-assignment schemes are inefficient when all nodes do not have data to send, since scarce resources are allocated to nodes that are not using them.

Random-access methods, on the other hand, do not assign users fixed resources. These are contention-based schemes, where nodes that have information to transmit must try to obtain bandwidth while minimizing collisions with other nodes' transmissions. These MAC protocols are more efficient than fixed-assignment MAC protocols when nodes have bursty data. However, they suffer from possible collisions of the data, as all nodes are contending for the resources. ALOHA is one of the earliest random access MAC protocols designed for wireless communications [34]. Although it is very simple, it is still used for certain applications. In ALOHA a node simply transmits a packet when it is generated. If the transmission of a packet is unsuccessful (*i.e.*, no acknowledgment is received), then it is retransmitted.

Often protocols use a hybrid approach, *e.g.*, combining TDMA and FDMA by allocating a certain time and frequency slot for each node. MAC protocols can be evaluated in terms of energy dissipation, fairness, and throughput, where the protocol is typically optimized to minimize energy dissipation, give each node its fair share of the bandwidth, and achieve high throughput [35, 36]. We present a detailed description of the MAC layer in Chapter 3.

### 2.2.5 Network Layer

In MANETs, often the source and the destination are beyond the direct transmission/reception ranges of each other. However, the lack of direct transmission/reception does not rule out the establishment of communication. Nodes that are not in direct communication range use multi-hop forwarding through peers. In fact, multi-hop forwarding is one of the most important advantages of MANETs. Such indirect communication through multi-hop forwarding is called routing, and it is the responsibility of the network layer to set up and maintain routes from a source to any desired destination. The network layer utilizes the MAC layer, which deals with single-hop data forwarding, to enable such end-to-end (multi-hop) data forwarding.

In order to provide efficient routes, the network layer needs sufficient information about the topology of the communication network. There are two main function of the network layer: route discovery and route maintenance. Route discovery is used to find a route from a source to a destination that meets some criteria (*e.g.*, shortest hop count, minimum cost), while route maintenance is used to maintain an existing route as the topology changes due to node mobility.

There are three basic types of routing based on the number of destinations: (*i*) unicasting, (*ii*) multicasting, and (*iii*) broadcasting (see Figure 2.7). In unicasting there is exactly one destination node. In multicasting there are multiple

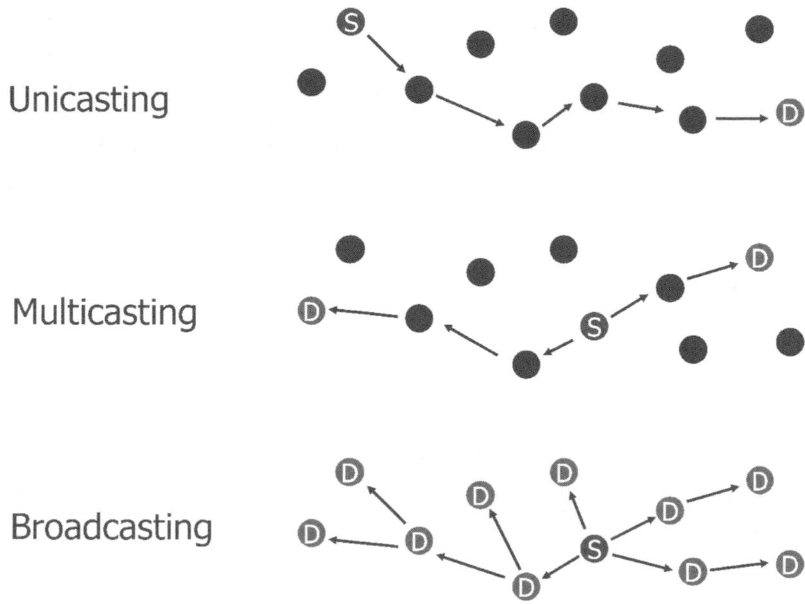


*Figure 2.7.* Illustration of unicast, multicast, and broadcast routing. S and D represent the source and destination nodes, respectively.

destination nodes. In broadcasting all the nodes in the network are destinations. We present a detailed description of the network layer in Chapter 4.

## 2.2.6 Transport Layer

The transport layer is responsible for end-to-end connection establishment, end-to-end data packet delivery, congestion control, and flow control. User Datagram Protocol (UDP) is a simple and connectionless transport layer protocol. The service provided by UDP for the application layer is to send encapsulated datagrams without having to establish a connection. The main advantage of UDP is its low overhead. UDP packet format is shown in Figure 2.8.

Real-Time Protocol (RTP) is a generic real-time transport protocol for multimedia applications. RTP is used in conjunction with UDP, thus two transport layer protocols are placed in the same layer. The basic function of RTP is to multiplex multiple real-time data steams onto a single stream of UDP packets. Each packet in an RTP stream is assigned a sequence number, which is incremented at each new packet. This allows the destination to detect missing data packets. Even if a packet is missing, it is not retransmitted because in real-time applications such as voice a late packet is of no use to the application. Thus, RTP does not provide flow control, error control, and packet recovery. Synchro-

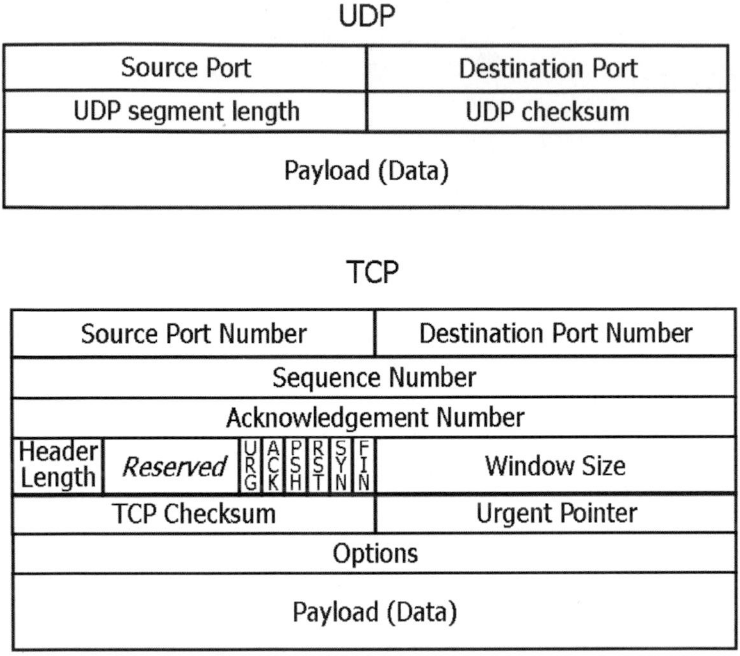

UDP

| Source Port | Destination Port |
|---|---|
| UDP segment length | UDP checksum |
| Payload (Data) | |

TCP

| Source Port Number | | | | | | | | | Destination Port Number |
|---|---|---|---|---|---|---|---|---|---|
| Sequence Number | | | | | | | | | |
| Acknowledgement Number | | | | | | | | | |
| Header Length | *Reserved* | URG | ACK | PSH | RST | SYN | FIN | | Window Size |
| TCP Checksum | | | | | | | | | Urgent Pointer |
| Options | | | | | | | | | |
| Payload (Data) | | | | | | | | | |

*Figure 2.8.* UDP and TCP packet formats.

nization of multiple parallel flows requires timestamping, which is provided by RTP. Real-Time transport Control Protocol (RTCP) is a protocol used to provide feedback on bandwidth, congestion, delay, and jitter to RTP, however, it does not transport data by itself. The feedback provided by RTCP can be used by the application layer to adapt to the network conditions [25].

UDP and RTP do not provide 100% data integrity (*i.e.*, some packets are lost between the source and destination), which is acceptable in real-time data communication (*e.g.*, voice communication). However, some applications require a high level of reliability and data integrity (*i.e.*, all packets must be transported to the destination). Transmission Control Protocol (TCP) is designed to serve these applications. TCP is a reliable, end-to-end, connection-oriented transport layer protocol. Through efficient feedback mechanisms it adapts to the varying congestion level of the network. Above all, it is the only protocol that provides a high level of data integrity to the application because the other layers below the transport layer cannot provide any such service (*e.g.*, any MAC layer protocol can provide at most a best effort service—it cannot guarantee that a particular packet is received). Especially in MANETs where the link qualities can fluctate unpredictably (handled by the physical and MAC layers) and the network topology can change rapidly (handled by the routing layer), the responsibility of the transport layer (TCP), which is end-to-end reliable data transportation, becomes extremely important. TCP packet format is shown in Figure 2.8.

### 2.2.7 Application Layer

The application layer is actually the only layer with which a user interacts. All the other layers are there to create a seamless interface for the networking needs of the application layer. Depending on the requirements of an application, the functionalities of the other layers change. For example, in data transfer packet delivery ratio should be 100% (transport layer packets), because packet loss is not tolerable. Thus, the transport protocol should be chosen as TCP. However, delay tolerance of data packets is not critical. On the other hand, in voice communications the important parameters are bounded packet delay and jitter, and some level of packet loss is tolerable. Thus, the RTP and UDP protocols should be used in time-critical applications, such as voice and video. Real-time voice is the application that we consider in this book. Chapter 5 further discusses the QoS issues for real-time voice.

Figure 2.9 illustrates streaming media communications in a multi-hop network. The application layer protocol is ITU G.711, which is a protocol for Pulse Code Modulation (PCM) [37]. It encodes voice with 8000 samples per second and eight bits per sample, which results in 64 kbps uncompressed speech [25]. The lower layers are transparent to the upper layers (*e.g.*, the application layer protocols at node A and at node D do not need to know about the existence of nodes B and C).

Although the design of a protocol using a layered approach enables the designer to separately design the different functions to achieve modularity [24], such an approach does not allow the separate layers to interact and therefore may not be optimal in all situations [26]. The alternative is to use a cross-layer design, which is discussed in the following section.

## 2.3 Cross-layer Design

It is argued in [26] that it is hard to achieve design goals such as energy efficiency and application-specific QoS requirements by using a system consisting of independently designed layers of the protocol stack. Alternatively, a cross-layer design that takes into account the specific QoS requirements of the application and tailors the rest of the protocol stack accordingly can achieve the design goals with much higher efficiency when compared to a general architecture [38].

Cross-layer design is a broad definition that includes various design alternatives. An extreme case for cross-layer design is collapsing the stack and

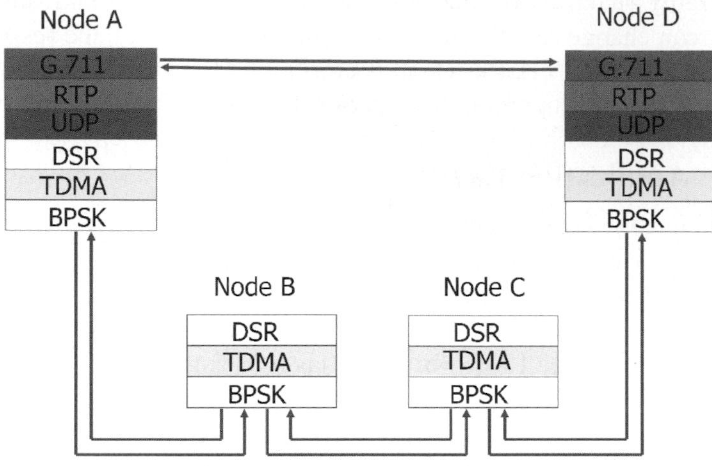


*Figure 2.9.* Audio communications through the network stack.

designing a completely integrated protocol architecture [39, 40]. Figure 2.10 shows two cross-layer design approaches. The first approach presented in the middle column shows a cross-layer design where the layers are kept intact but all the layers share information. The second approach presented in the right column illustrates the merging of the application and transport layers into a single layer and the merging of the network and MAC layers into a single layer.

To illustrate the improvements that can be achieved by a cross-layer design that enables information sharing among different layers, we will give a cross-layer design example taken from the TRACE protocols [41–51]. The amount of information a node can receive in a single-hop broadcast medium may be higher than the usable range of the node (*i.e.*, the application layer), in which case the node should select to receive only certain data packets. For example, if the number of simultaneous conversations in a group of people, communicating through a single-hop broadcast network, exceeds a certain threshold, then each user should select a subset of the voice packets based on some discrimination criteria like proximity, and discard the rest of the packets. The straightforward approach, which is receiving all data transmissions, keeping the ones desired, and discarding the others, is an inefficient way of discriminating data. However, in an independently designed protocol stack there is no other way of discriminating the data packets, because the lower layers (*e.g.*, the MAC layer) are not aware of the requirements of the application layer. An energy-efficient method for discriminating data is information summarization prior to data transmission [52], which can be performed via MAC packets if the application and MAC layers have means for information sharing, which necessitates a cross-layer design.

It has been shown that network protocols can be defined on an application-specific basis, where protocols are created by an application to support the functions it requires [38, 53, 54]. The LEACH protocol architecture [38] for wireless sensor networks employs the technique of cross-layer design to expose lower layers of the protocol stack to the requirements of the sensor network application. The results reported in [38] illustrate the high performance that can be achieved despite the harsh conditions of the wireless channel using an application-specific architecture.

The protocol architectures described in [55–58] use a cross-layer design to expose the topology/capacity changes due to congestion, channel errors, or mobility throughout the different layers. Thus, the burden of coping with these problems are not handled by a single layer; instead, several layers take counter measures to compensate for the adverse affects of the environment with greater efficiency.

Application-specific data routing protocols described in [52, 59] use a cross-layer design by creating an application layer aware network layer to achieve data centric routing. The results presented in these studies have shown that the

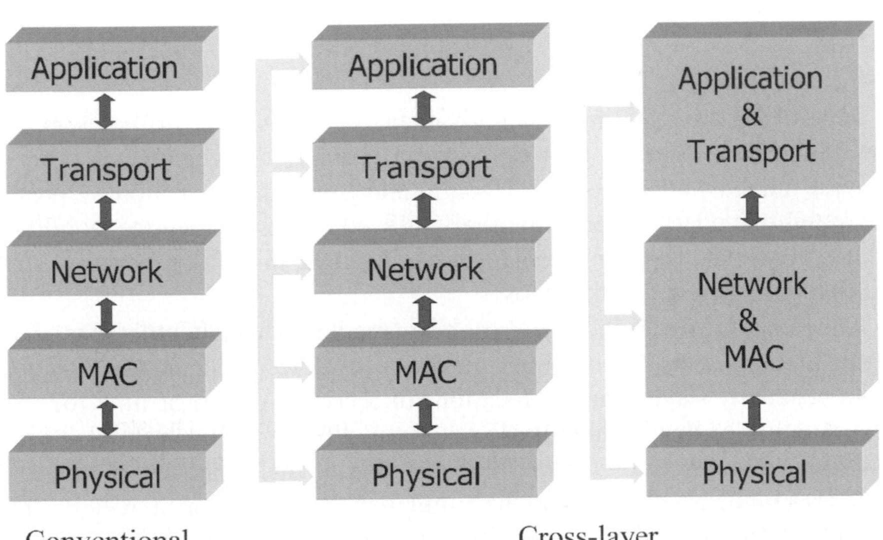


*Figure 2.10.* The left column shows a conventional layered protocol stack. The middle column shows a cross-layer design, where layers share information while keeping the layers intact. The right column shows another cross-layer design where application and transport layers are combined into a single entity and network and MAC layers are merged.

close interaction and integration between different layers of the protocol stack lead to great performance improvements when compared to a relatively blind layering approach.

Cross-layer design is becoming an integral part of several developing wireless standards [60]. 3G standards such as CDMA2000, Broadband Radio Access Network (BRAN) of HiperLAN2, High Speed Downlink Packet Access (HS-DPA) of 3G Partnership Project, and IEEE Study Group on Mobile Broadband Wireless Access Networks are some of the large scale design efforts that use a cross-layer design [61].

The TRACE family of protocol architectures is designed by using a cross-layer design approach. The MAC layer in SH-TRACE is designed specifically for voice communications (Chapter 6). MH-TRACE inherits the application-specific cross-layer design of SH-TRACE and extends it to multi-hop networks (Chapter 7). NB-TRACE and MC-TRACE extend MH-TRACE for network-wide broadcasting and multicasting by merging the MAC layer and network layer (Chapter 10 and Chapter 11).

Both cross-layer and independently layered protocol architectures have their advantages and disadvantages. However, for the sake of explaining various concepts of wireless networks in a concise fashion, it is better to use an abstraction by analyzing the spectrum of functionalities of a network within an organization of independent layers of a conventional protocol stack. Protocol architectures presented in this book are mostly related with the MAC and network layers, which we will discuss in detail in Chapters 3 - 5.

## 2.4 Mobility

Mobility is the most important advantage of a MANET for the users. On the other hand, it is the most challenging problem to overcome in order to provide reliable high performance communications. In this section we present the mobility models that are commonly utilized for modeling node mobility in ad hoc networks. These models are also utilized for mobility generation in the simulations discussed in this book.

The Random Way Point (RWP) mobility model is a widely utilized model to create mobility scenarios in mobile ad hoc network research. In RWP a mobile node begins by staying in one location for a certain period of time [62, 63]. The duration of this immobility is determined by a random variable, which is called the pause time. Upon the expiration of this time, the mobile node chooses a random destination in the simulation area and a speed, which is also a random variable. The mobile node then travels toward the newly chosen destination at the selected speed. Once the node arrives at its destination, the process repeats itself. As reported in [64, 65], the RWP mobility model does not produce uniform node distribution in the network. Instead, the node density at the center is higher than the node density at the other parts of the network

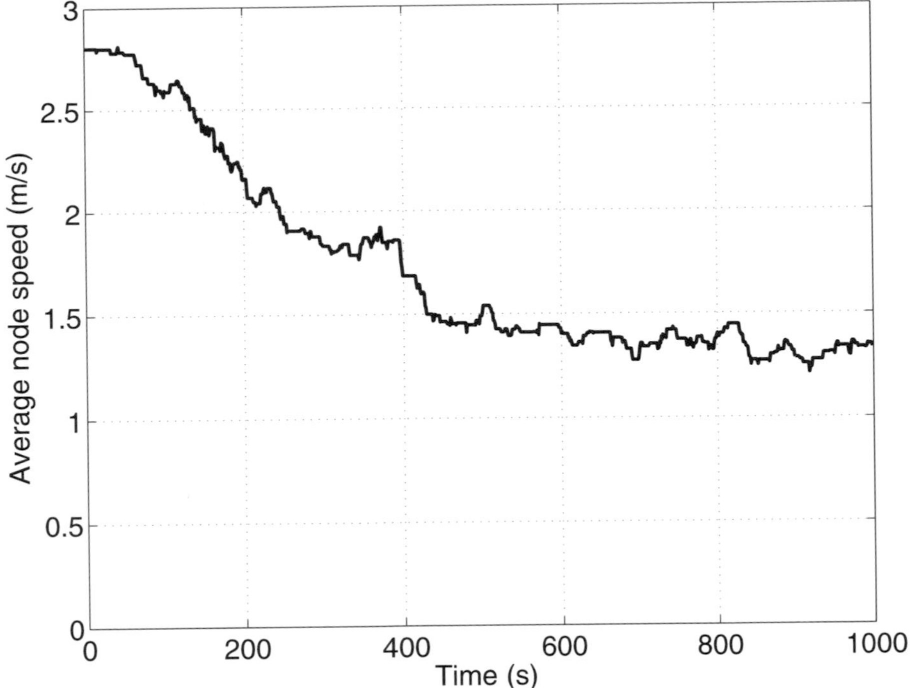


*Figure 2.11.* Average node speed for a simulation scenario created by the random waypoint mobility model with 80 nodes over 1 km by 1 km area. The node speeds are chosen randomly from [0, 5 m/s] with zero pause time.

[65]. Furthermore, the average instantaneous node speed is shown to decrease over time [64]. Throughout this book we used the RWP mobility model for the generation of mobility scenarios in multi-hop networks. The node speeds are chosen from a uniform random distribution between 0.0 m/s and 5.0 m/s (the average pace of a marathon runner). The pause time is set to zero to avoid non-moving nodes throughout the simulation time. Figure 2.11 shows the average node speed as a function of time. Average node speed at the beginning of the scenario is about 2.7 m/s, but at the end of the scenario (1000 s) the average node speed decreases to 1.3 m/s. Figure 2.12 presents the node distribution at 1000 s.

Reference Point Group Mobility (RPGM) is a commonly used mobility model for the simulation of coordinated group mobility scenarios [66]. We used a coordinated hierarchical mobility model in our single-hop mobility simulations (Chapter 6), which is called Hierarchical Reference Point Group Mobility (HRPGM). This model is similar to the RPGM model. In our HRPGM model, nodes are moving around a global center randomly, from which they cannot be farther than a radius of $r_g$. The global center is also mobile, and its motion can

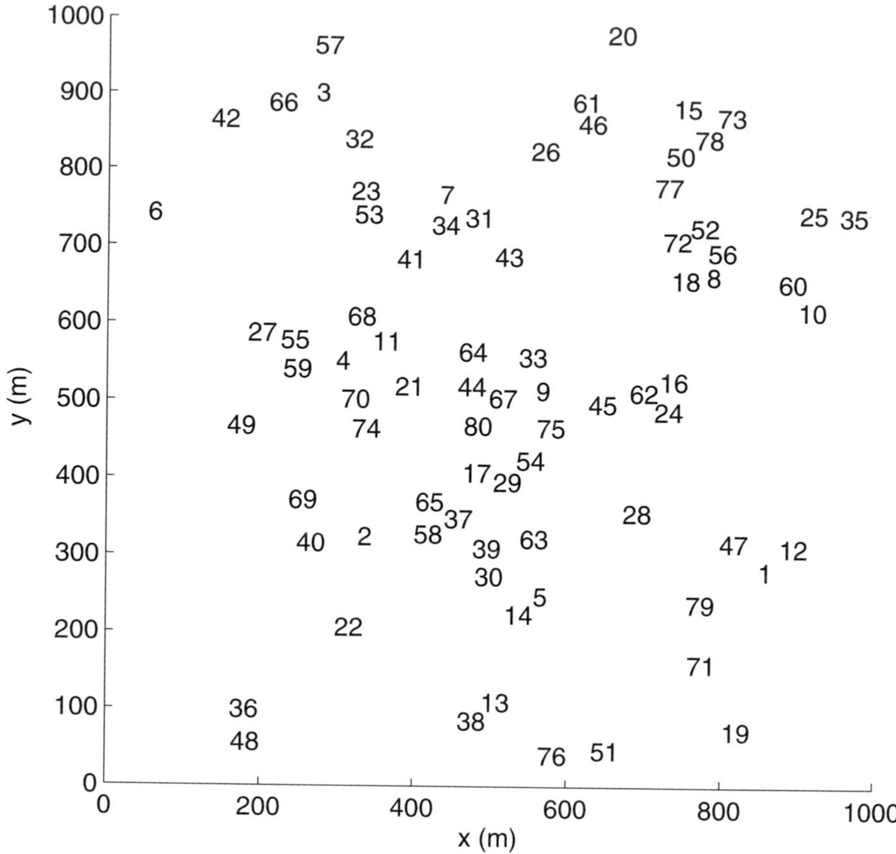


*Figure 2.12.* Node distribution produced at 1000 s of the mobility scenario by the random waypoint mobility model with 80 nodes over 1 km by 1 km area. The node speeds are chosen randomly from [0, 5 m/s] with zero pause time.

follow an arbitrary motion pattern. It is possible to use the Random Way Point mobility model (RWP) [62] to create the motion pattern of the global centers. In addition, nodes are further divided into sub-clusters within the global cluster. Each sub-cluster has its own local center, and the members of the sub-clusters lie inside a circle with a radius of $r_l$ and centered at the sub-cluster center. Local centers are also moving randomly without leaving the large circular area around the global center. Actually, each node follows a mobility pattern as if it was generated by the RWP model with two level hierarchical constraints, which are not leaving the global circle centered at the global center and not leaving the local circle centered at the local center. In order to allow more flexibility in the motion model, we expanded the basic mobility pattern by introducing the "bunching" and "spread-out" modes to our model. Bunching means nodes are very close to each other and there are no sub-clusters. Spread-out is the basic

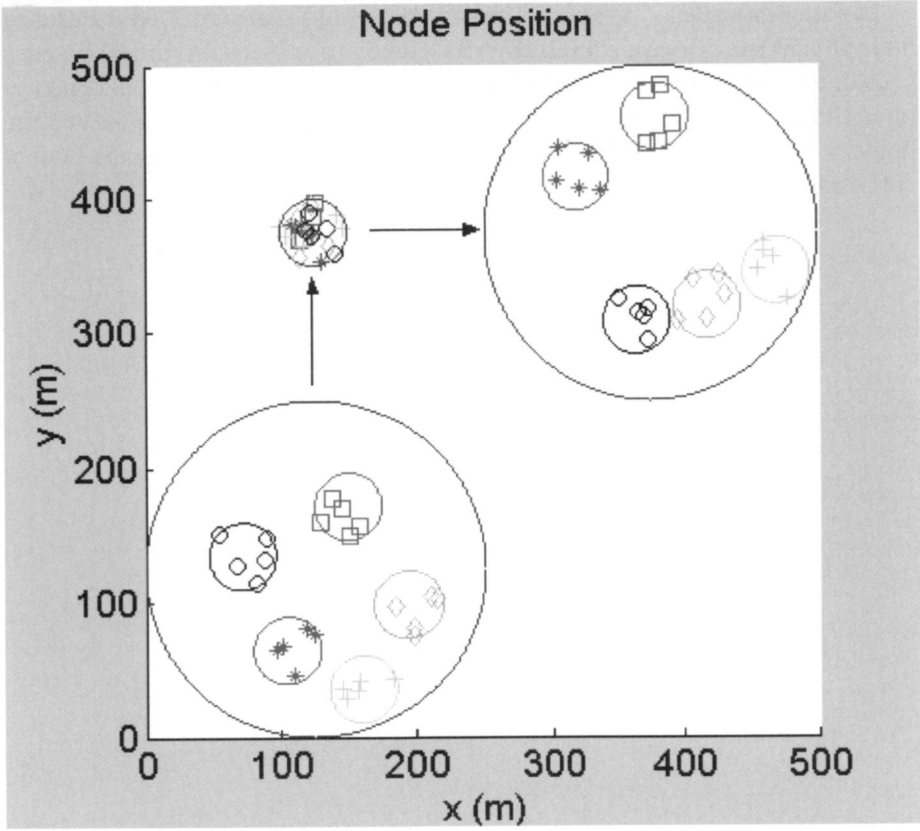


*Figure 2.13.* Combined snapshots of node positions in time plotted over a 500 m by 500 m grid. The lower-left corner of the figure is the snapshot at time 0.0 s. The upper-left corner shows the nodes in bunching mode at 50.0 s. The final position of the nodes at 100.0 s is in the upper-right corner of the figure.

mobility scheme, where sub-clusters are spaced to avoid intersections between them.

The global center moves with an average speed of 5 m/s, which is fairly high for such a tightly coordinated mobility pattern; yet it is realistic for high-pace events, like military operations, search and rescue operations, and disaster recovery operations. The average speed of both the individual nodes and the sub-cluster centers is also 5 m/s. The global radius for the global cluster, $r_g$, the local radius of sub-clusters, $r_l$, and the radius in bunching mode, $r_b$, are 125 m, 25 m, and 25 m, respectively. The minimum inter-sub-cluster distance is 50 m, and the minimum distance between the nodes is 4 m in spread-out mode and 1 m in bunching mode.

The mobility scenario for 25 nodes is shown in Figure 2.13 over a grid of 500 m by 500 m. There are 5 sub-clusters with 5 nodes, each. At time 0 s, nodes start in the spread-out mode in the lower-left corner, with the global center at ($x$ = 125 m, $y$ = 125 m). At time 50 s, nodes complete bunching around the point ($x$ = 125 m, $y$ = 375 m). The scenario ends with the final spread out at 100 s with the global center at ($x$ = 375 m, $y$ = 375 m).

# Chapter 3

# MEDIUM ACCESS CONTROL

In wireless communications, the channel, which is the common interface that connects the nodes, is a shared resource. Thus, access to this shared resource needs to be coordinated either centrally or in a distributed fashion. The objective of controlled access is to avoid or minimize simultaneous transmission attempts (that will result in collisions) while maintaining a stable and efficient operating region for the whole network [25, 27, 32, 67]. Medium Access Control (MAC) protocols are responsible for the access control.

Collisions occur if multiple nodes transmit at the same time within range of a given destination, and the destination's receiver cannot resolve the composite signal created due to this uncontrolled superposition in favor of any of the senders. However, if one of the components of the composite signal is dominant to the other, then the destination node receives the high power signal and the other transmissions are not heard (see Figure 3.1). This phenomenon is known as capture.

MAC protocols can be classified into two categories based on the assignment of channel access: (*i*) fixed assignment and (*ii*) random access, which will be discussed in the following subsections.

## 3.1 Fixed Assignment MAC Protocols

The straightforward solution for medium access is fixed assignment of the resources to the users through Frequency Division Multiple Access (FDMA), Time Division Multiple Access (TDMA), and Code Division Multiple Access (CDMA), which are illustrated in Figure 3.2. Generally, fixed assignment schemes are associated with a base station that assigns the resources. Data transmission is either from the base station to the ordinary nodes (downlink) or from the ordinary nodes to the base station (uplink). Direct peer-to-peer

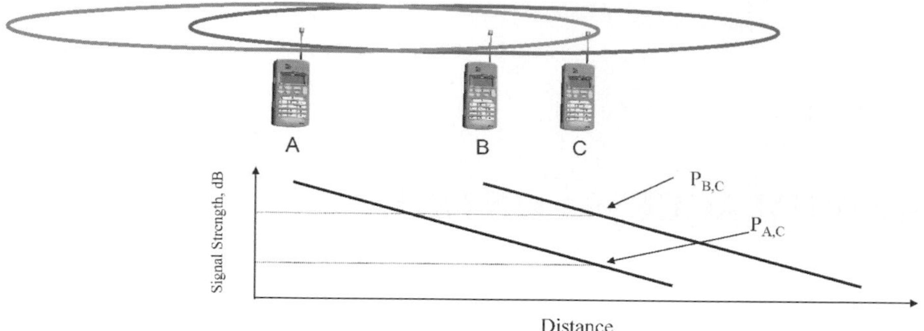


*Figure 3.1.* Node B is closer to node C than node A. Simultaneous transmissions by node A and node B do not result in collisions because the signal strength of the transmission by node B at node C's receiver ($P_{B,C}$) is much higher than that of node A ($P_{A,C}$). This effect is known as "capture".

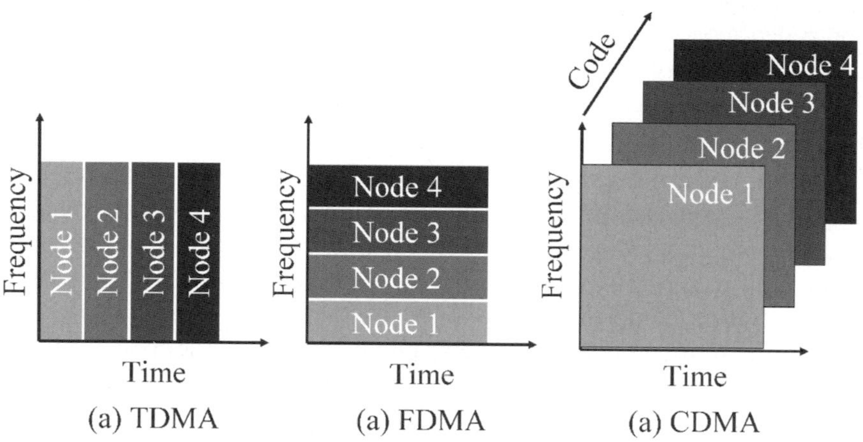


*Figure 3.2.* Fixed assignment medium access control protocols: (a) Time Division Multiple Access (TDMA), (b) Frequency Division Multiple Access (FDMA), (c) Code Division Multiple Access (CDMA).

communication is not supported. Instead, the base station relays the information, ev en if the nodes are in direct transmission range of each other.

In TDMA time is organized into cyclic frames, and each frame consists of a fixed number of slots. Each node is assigned a fixed slot within the frame to transmit its data (see Figure 3.2 (a)). During this time-slot, no other node should access the channel, so there are no collisions and the throughput is equal to the total data transmitted by each node. Assume there are $N$ nodes that have data to transmit and the time for each frame is $t_f$ and the channel bandwidth is

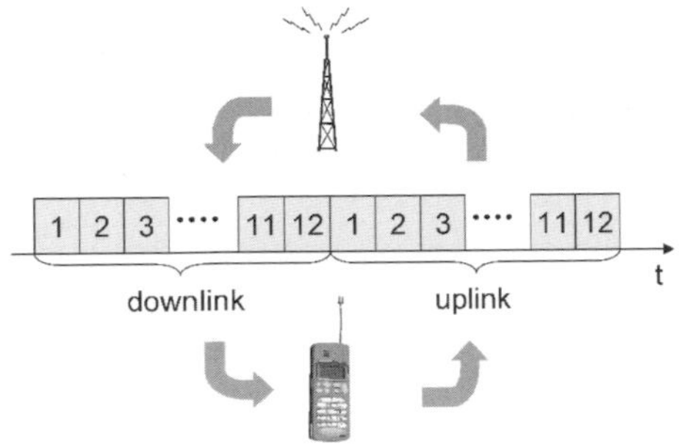


*Figure 3.3.* Digital European Cordless Telephone (DECT) uses TDMA as the MAC layer. The frame length is 10 ms consisting of 24 time slots of duration 417 s, of which 12 are used for downlink (*i.e.*, from the base station to the mobile nodes) and 12 are used for uplink (*i.e.*, from the mobile nodes to the base station).

$B_w$. Each node will get $t_s = \frac{t_f}{N}$ seconds in which to transmit data. Assuming a 1 bit/sec/Hz signaling scheme, each node can transmit $B_w t_s = B_w \frac{t_f}{N}$ bits per frame or $R_b = \frac{B_w}{N}$ bps.

Transmission occurs in bursts in TDMA, which reduces energy dissipation compared to a protocol where the transmission is continuous because the transmitter hardware (*e.g.*, the phase-locked loops for frequency generation, the power amplifier) can be turned off when the node is not transmitting. In addition, this is a simple protocol to implement in the radio hardware. However, this protocol requires that some node have knowledge of all the transmitting nodes to create the schedule, and if the number of nodes that need to transmit data is variable, the schedule must be changed often, adding significant overhead to the protocol. In addition, using TDMA requires that all nodes be time-synchronized and requires guard slots to separate users [27]. These extra bits add to the overhead of a TDMA system. Tight synchronization is necessary in TDMA schemes to avoid overlapping transmissions [68].

Digital European Cordless Telephone (DECT) uses TDMA as the MAC layer (see Figure 3.3). The frame length is 10 ms consisting of 24 time slots of duration 417 s, of which 12 are used for downlink (*i.e.*, from the base station to the mobile nodes) and 12 are used for uplink (*i.e.*, from the mobile nodes to the base station) [32].

In FDMA, the bandwidth is divided into slices such that each user gets a unique section of bandwidth in which to transmit data (see Figure 3.2 (b)). Because no node is supposed to transmit in the bandwidth slice given to another node, there are no collisions between nodes' data. If there are $N$ nodes that must share the $B_w$ bandwidth, each node gets a frequency slice of size $\frac{B_w}{N}$ in which to transmit. As expected, given the same bandwidth $B_w$, the same number of

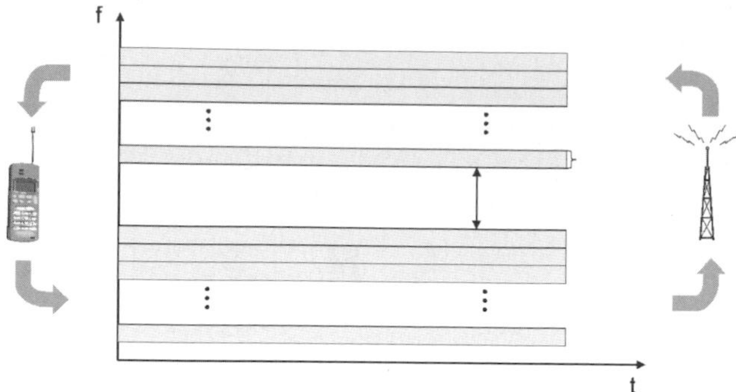


*Figure 3.4.* Global System for Mobile communication (GSM) uses FDMA as the MAC layer. The frequency band is divided into 256 channels (128 channels for uplink and 128 channels for downlink), and the carriers are separated by 200 kHz.

users $N$, and the same signaling scheme (1 bit/sec/Hz), both FDMA and TDMA give each user the same bitrate ($R_b = \frac{B_w}{N}$ bps) under ideal conditions.

Transmission is continuous in FDMA, reducing the number of guard and synchronization bits needed compared to TDMA, thereby decreasing overhead (although FDMA may require guard bands to ensure transmissions do not overlap in frequency). However, FDMA requires that the transmitter hardware be on at all times, increasing the energy dissipation compared with a burst transmission protocol such as TDMA. In addition, FDMA requires good filtering to ensure that energy transmitted in the neighboring slices of bandwidth do not interfere with the transmission. FDMA also requires that some node have knowledge of all the transmitting nodes to allocate bandwidth appropriately.

Global System for Mobile communication (GSM) uses FDMA as the MAC layer (see Figure 3.4). The frequency band is divided into 256 channels (128 channels for uplink and 128 channels for downlink), and the carriers are separated by 200 kHz [32].

In both TDMA and FDMA systems, the bandwidth is pre-allocated (separated in time or frequency for each user). The advantage of pre-allocation of the limited resources is that no collisions will occur, since each node has a unique time/frequency slice of bandwidth in which to transmit its data. The disadvantage of pre-allocation is that if nodes do not have data to send, the resources allocated to that node are wasted. To avoid waste, schedules can be changed often, reallocating time or frequency as needed. The problem with this is that it adds significant overhead to the protocols, as the controlling node must poll all the nodes to find out which ones have data to transmit, it must appropriately allocate resources, and it must transmit the schedule to all nodes.

CDMA is more elegant than both TDMA and FDMA, because the orthogonality required for the separation of transmissions is achieved by code division, which is synthesized by using both time and frequency (see Figure 3.2 (c)). In Direct Sequence Spread Spectrum (DSSS), the CDMA signals are spread into a larger frequency band than the signal bandwidth. Using DSSS, several nodes can transmit at the same time using the same bandwidth by spreading their data using a unique spreading code (typically a pseudo-random noise sequence). Reception of the signal is done by correlating with the spreading code used to transmit that signal. All other signals will appear as white noise after decorrelation with the correct spreading code. Power control, where each node sets its transmit power to ensure the same amount of power at the receiving node, is very important in CDMA systems. If all nodes transmit at the same power, signals from nodes close to the receiving node will drown out signals from nodes further away from the receiving node. This is known as the *near-far* problem. In addition to reducing interference among different signals, using power control minimizes energy dissipation at the nodes.

As more nodes transmit data using their unique spreading code, the Signal-to-Noise Ratio (SNR) of each transmission is reduced, whereas if fewer nodes transmit data, the SNR of each transmission is increased. Therefore, the performance degrades slowly as the number of nodes is increased [27]. The SNR will depend on the amount of spreading and the number of interfering signals. The number of simultaneous transmissions in a CDMA system is [69]:

$$N = \eta_b c_d \times \frac{B_w}{R_b(\frac{E_b}{I_o})} \tag{3.1}$$

where $\eta_b$ is the bandwidth efficiency factor, $c_d$ is the capacity degradation factor due to imperfect automatic gain control, $B_w$ is the total bandwidth, $R_b$ is the information rate, and $\frac{E_b}{I_o}$ is the bit energy to interference ratio required to achieve an acceptable probability of error. Assuming ideal conditions, this equation simplifies to:

$$N = \frac{B_w}{R_b \frac{E_b}{I_o}} \tag{3.2}$$

The minimum SNR required ($\frac{E_b}{I_o} = 1 = 0$ dB) is achieved when the minimum spreading of the data is performed. In this case, $R_b = \frac{B_w}{N}$. Therefore, under ideal conditions with minimum spreading, each of the $N$ users can transmit at the same bitrate as in the TDMA and FDMA systems.

Ideally it is possible to have an infinite number of orthogonal spreading codes, but the number of available fixed length spreading codes is limited. For example, the number of spreading codes (called Barker codes) are limited to seven in IEEE 802.11 [2].

In Frequency Hopping Spread Spectrum (FHSS) the CDMA signal is modulated into different frequencies in a fast manner (*i.e.*, frequency hopping). The hopping pattern is a pseudo random sequence, which is agreed upon by the

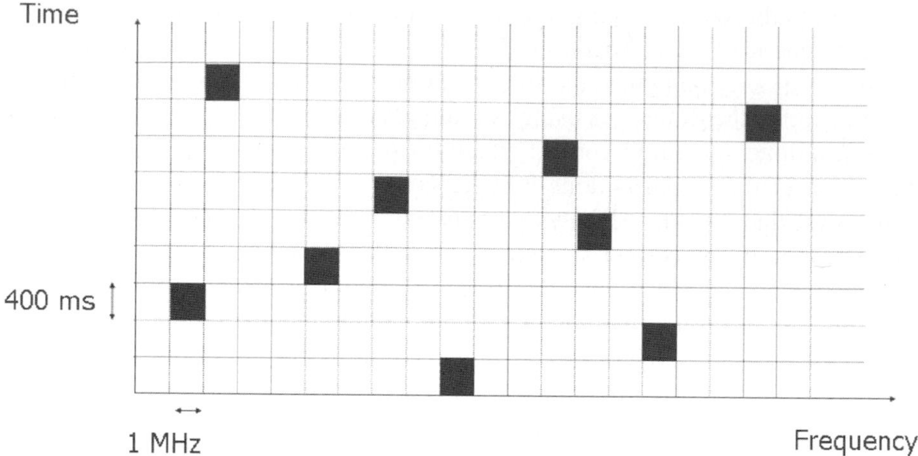


*Figure 3.5.* Illustration of Frequency Hopping Spread Spectrum (FHSS).

transmitter and receiver (see Figure 3.5). All the other nodes, which do not know the hopping pattern, observe the modulated signal as bursty noise.

Some MAC protocols exploit the use of multiple transmit and receive antennas to increase system capacity using Space Division Multiple Access (SDMA). Smart antennas are antenna arrays that use beamforming techniques to point the receiver towards a particular transmitter while nulling out other transmitters and the effects of multipath from all the transmitters [70]. Thus the receiver is no longer omni-directional but is capable of selecting the direction of the desired transmitter. Another SDMA protocol, the Bell Labs Adaptive Space-Time (BLAST) protocol, uses antenna arrays at both the transmitter and the receiver to point not only the receiver but also the transmitter [71]. The drawback of these systems is the high complexity and cost of having multiple transmit and/or receive antennas.

Although fixed assignment schemes completely eliminate collisions through pre-allocation of the resources, this advantage comes with a sacrifice, which is wasting bandwidth due to underutilization. This is because, in many applications, most of the time, nodes do not have data to send.

The alternative of fixed assignment schemes is random access. These are contention-based schemes, where nodes that have information to transmit must try to obtain bandwidth while minimizing collisions with other nodes' transmissions. These MAC protocols are more efficient than fixed assignment schemes when nodes do not have continuous data. However, random access protocols suffer from collisions due to the randomness of the channel access, and they have stability problems due to their dynamic nature. Nevertheless, almost all MAC protocols used for MANETs are based on the random access principle.

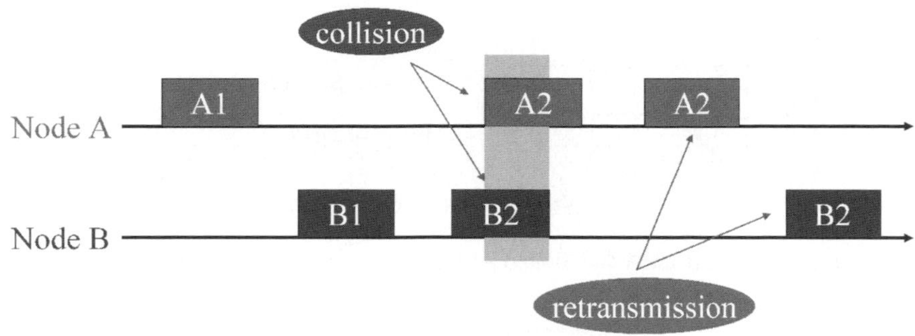


*Figure 3.6.* ALOHA medium access.

## 3.2 Random Access MAC Protocols

To illustrate the operation of random access protocols we will start with the first and simplest MAC protocol - ALOHA [34]. This protocol derives its name from the ALOHA system, a communications network developed at the University of Hawaii to enable wireless communication among the campuses located at different Hawaiian islands and first put into operation in 1971.

The design principle of ALOHA is very simple: whenever a node has data to send it is transmitted right away. If the transmitting node is the only transmitter in the medium, then the packet transmission is successful. For example, in Figure 3.6, packet A1 transmitted by node A and packet B1 transmitted by node B are successfully received by the base station. A transmitting node knows that its data transmission is successful by the reception of an acknowledgment (ACK) packet transmitted by the base station in response to the data packet. Due to the random nature of the channel access, packet collisions are unavoidable, like the collision of packets A2 and B2, shown in Figure 3.6. Since neither node A nor node B receives an ACK packet from the base station, they know that their packet transmissions are not successful and they retransmit after waiting a random amount of time to avoid successive collisions.

The vulnerable period for a packet transmission, where any transmission attempt by any other node will result in a collision, is two packet lengths due to the lack of synchronization. This is a factor that limits the maximum throughput achievable by the ALOHA protocol, which is 18.4% (see Figure 3.7).

In Slotted ALOHA (S-ALOHA) [72] time is divided into slots and nodes can start their packet transmissions only in the beginning of each slot (see Figure 3.8), which requires global clock synchronization and reduces the vulnerable period to one packet time. Due to the reduction in the packet vulnerable time, the maximum throughput of S-ALOHA is double the maximum throughput achievable by ALOHA, or 36.8% (see Figure 3.7). Although the throughput

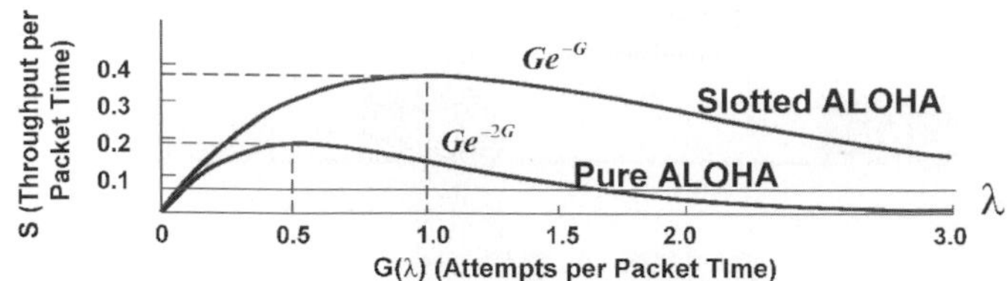


*Figure 3.7.* ALOHA and Slotted ALOHA throughput versus offered load.

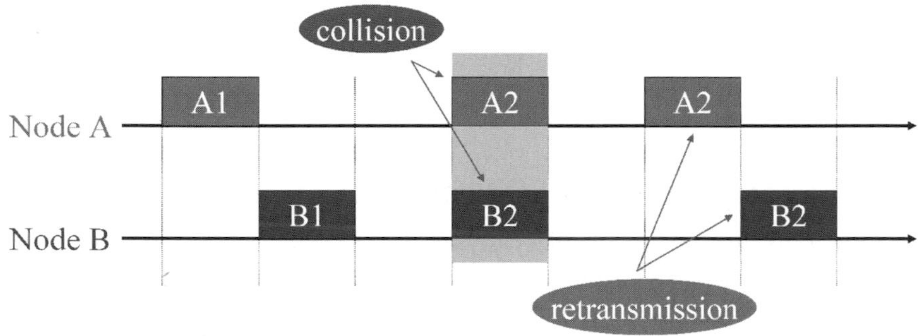


*Figure 3.8.* Slotted ALOHA medium access.

efficiency of S-ALOHA is low, it is still being used in applications like satellite communications where transmission delays are long [32].

Stability is an important problem in ALOHA and S-ALOHA, which may degrade system performance significantly [73]. For example, if the offered load exceeds the optimal operating point (*i.e.*, 50% and 100% of the effective bandwidth for ALOHA and S-ALOHA, respectively), throughput starts to decrease, eventually reaching zero throughput due to the excessive collisions.

The ALOHA schemes do not make use of channel feedback information, which is the main reason for their relative inefficiency. It is possible to achieve better throughput if the channel is listened to before transmitting. For example, if node A (see Figure 3.6) had listened to the medium before transmitting packet A2, it would have heard the ongoing transmission of node B (packet B2) and deferred its transmission until packet B2's transmission was completed. Thus, an obvious collision could have been avoided. Carrier Sense Multiple Access (CSMA) protocols use channel feedback (listen-before-talk) to achieve high throughput efficiency [25]. There are basically three versions of CSMA: (*i*) 1-persistent CSMA, (*ii*) non-persistent CSMA, and (*iii*) $p$-persistent CSMA.

In 1-persistent CSMA a node listens to the medium before transmitting its packet. If the medium is busy, transmission is differed until the channel is sensed idle. Due to the use of additional information, the throughput of 1-persistent CSMA is better than that of the ALOHA schemes (see Figure 3.9). However,

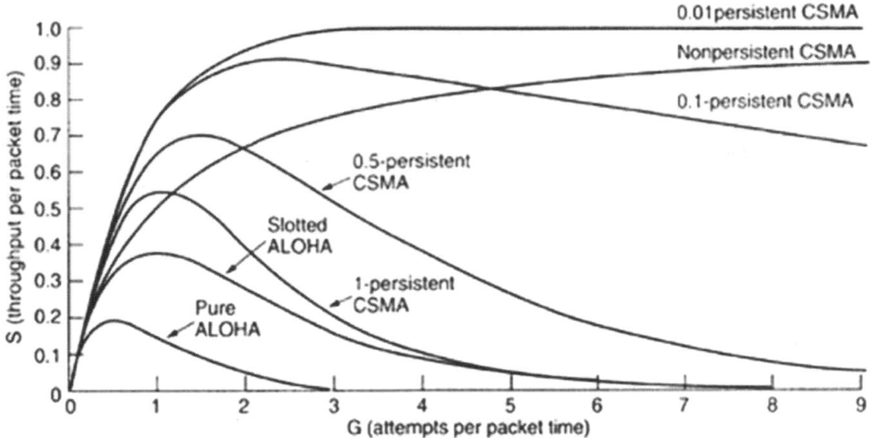


*Figure 3.9.* Comparison of the throughput efficiency versus offered load for the ALOHA and CSMA schemes. The propagation delay is small when compared to the packet length.

in the case of multiple nodes deferring simultaneously, packet collisions are unavoidable, because all of them will transmit their packets at the same time upon the completion of the ongoing transmission.

In non-persistent CSMA, a node defers for a random time if the channel is sensed busy. At the end of the defer time the channel is sensed again, and if the channel is idle the node transmits its packet; otherwise, the node continues to defer. Non-persistent CSMA eliminates most of the collisions that would result from multiple users transmitting simultaneously upon sensing the transition from busy to idle in 1-persistent CSMA.

$p$-persistent CSMA is a generalization of the 1-persistent CSMA scheme. If the channel is sensed busy, the node defers until the medium becomes idle. When the channel is sensed idle, the node transmits with a probability $p$. With a probability $q = 1 - p$ the node defers for one slot time. If that slot is idle, the node transmits with a probability $p$ or defers again with probability $q$.

Figure 3.9 shows that the CSMA schemes outperform the ALOHA schemes when the propagation delay is short compared to packet length. However, for longer propagation delays, CSMA protocols become relatively inefficient when compared to ALOHA schemes, due to the inability of the nodes to accurately sense the channel. Nevertheless, in most MANET scenarios propagation time is negligible when compared to the packet length.

Random access methods can further be classified into two categories: (*i*) centralized and (*ii*) distributed. Centralized MAC protocols are generally used in single-hop networks due to the availability of the global network status in each node. On the other hand distributed MAC protocols are used for multihop

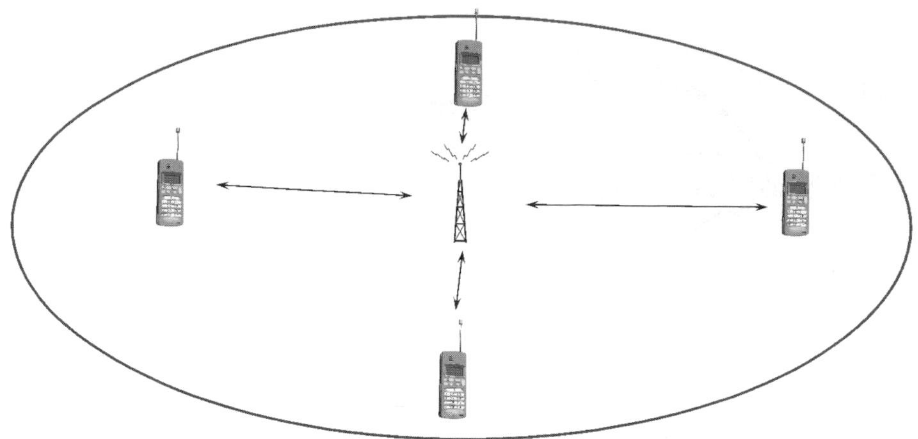

*Figure 3.10.* Star topology network—the base station is in the center.

networks, where monitoring and conveying the global network status is not feasible.

In a distributed MAC protocol, radios communicate without a central controller or base station. In other words, every radio should create its own access to the medium through a predetermined set of rules (*e.g.*, IEEE 802.11 [8]). A centralized MAC protocol, on the other hand, has a controller node or a base station that is the maestro of the network (*e.g.*, Bluetooth [3] or IEEE 802.11 in infrastructure mode). All the nodes in the network access the medium through some kind of schedule determined by the controller.

In the following section we will discuss centralized MAC protocols.

### 3.2.1 Centralized MAC Protocols

Centralized MAC protocols are designed to operate in single-hop networks. There are two possible topologies for a single-hop network. The first one is the star topology, where the base station is in the center of the network and the other nodes are in the one-hop neighborhood of the base station (see Figure 3.10). In this topology all traffic flows through the base station. The second topology is the fully connected single-hop topology, where all the nodes are in the single-hop neighborhood of each other (see Figure 3.11).

Centralized MAC protocols are generally more deterministic than distributed MAC protocols, which is a desirable feature for real-time traffic with delay constraints. As a result, it is advantageous to use a centralized MAC protocol in a single-hop network that supports real-time traffic delivery. For example, a distributed MAC protocol such as IEEE 802.11 cannot guarantee bandwidth or delay constraints or fair medium access. In fact, all of these parameters are

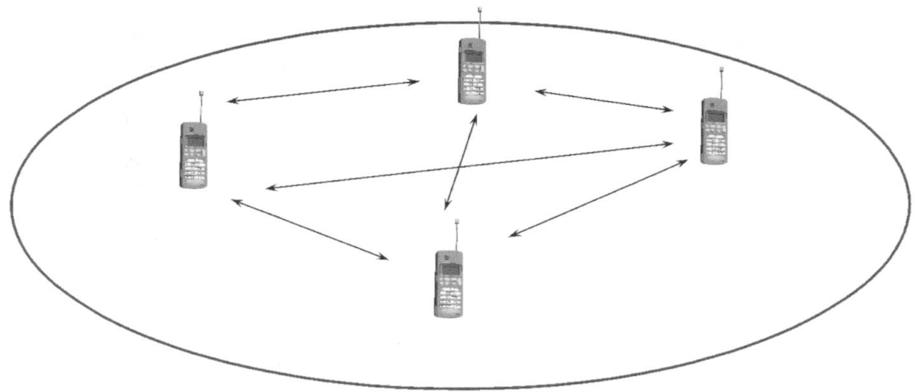

*Figure 3.11.* Fully connected single-hop wireless network.

functions of the data traffic, and they become unpredictable and often unacceptable at high data rates [74]. However, some centralized algorithms (*e.g.*, PRMA [6, 7]) can guarantee some of the above requirements within certain ranges by making use of coordination via scheduling [35]. Furthermore, when using a distributed MAC protocol such as IEEE 802.11, all nodes should be active all the time, because they do not know when the next transmission is going to take place [21]. However, using a centralized MAC protocol such as Bluetooth, nodes can enter sleep mode frequently due to the explicit polling of the slave nodes by the master node, which is an effective method to save power.

In a centralized MAC protocol, the two most important issues are the controller assignment and the data transmission schedule, which correspond to the coordinator and the coordination, respectively. The coordinator could be a fixed predetermined radio that is the sole controller for the entire network lifetime. The main drawback of this approach is that whenever the controller dies, the whole network also dies. The controller dissipates more energy than other nodes because of its additional processes and transmissions/receptions. Because of this higher energy dissipation, most possibly the controller will run out of energy before all the other nodes, leaving the entire network inoperable for the rest of the network lifetime, even though many other remaining nodes have enough energy to carry on transmissions/receptions.

The data transmission schedule could also be fixed, but this does not allow the system to adapt to dynamic environments such as nodes entering the network. The alternative approach to a fixed controller and schedule is dynamic controller switching and schedule updating, which is a remedy for the problems described above. However, this approach comes with its own problems: overhead in controller handover and increased overhead in the schedule updates.

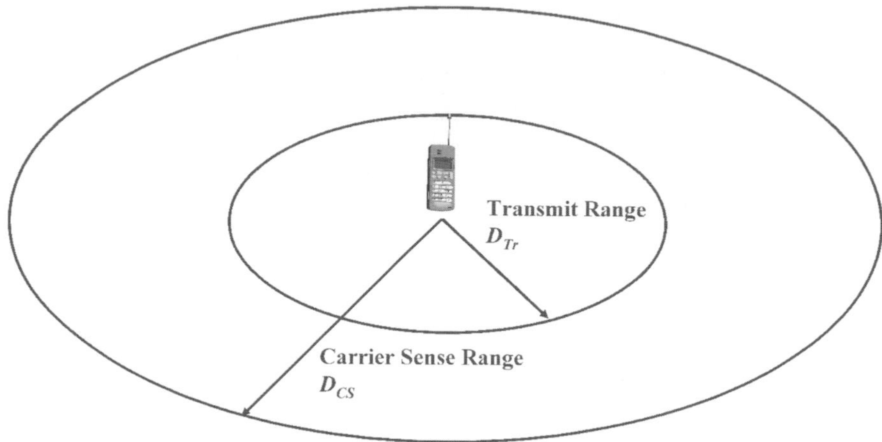


*Figure 3.12.* Illustration of transmit and carrier sense regions.

Although centralized MAC protocols are better than distributed protocols in single-hop networks, in multi-hop networks centralized control is not practical. Distributed MAC protocols, which will be discussed in the following subsection, are the only practical alternative for multi-hop networks.

### 3.2.2 Distributed MAC Protocols

From the perspective of the MAC layer, the network is a two-hop radius disk. The first hop is the direct reception range, where direct communication is possible. Although the second hop is not in a node's direct reception range, it is in the node's physical carrier sense range (see Figure 3.12), which means that direct communication is not possible due to the low signal strength but it is still possible to sense a busy medium (*i.e.*, a two-hop neighbor is in the carrier sense range, where, on the average, it is not possible to correctly detect if the transmitted bit is a one or a zero, but it is possible to distinguish the transmission from the background noise). The physical carrier sense range mainly depends on the sensitivity of the receiver and the radio propagation characteristics.

To avoid collisions, nodes create a temporary (per packet) coordination for each packet transmission. There have been many MAC algorithms to avoid collisions and coordinate the channel access in a distributed and per packet basis [23]. We will discuss several representative algorithms to sample the literature on distributed MAC protocols.

Busy Tone Multiple Access (BTMA) [128] is an example of a MAC protocol that uses out-of-band-busy-tone signals to prevent hidden nodes. Hidden nodes are nodes that are not in the transmission range of each other but their transmissions create collisions at the destination (see Figure 3.13). In BTMA,

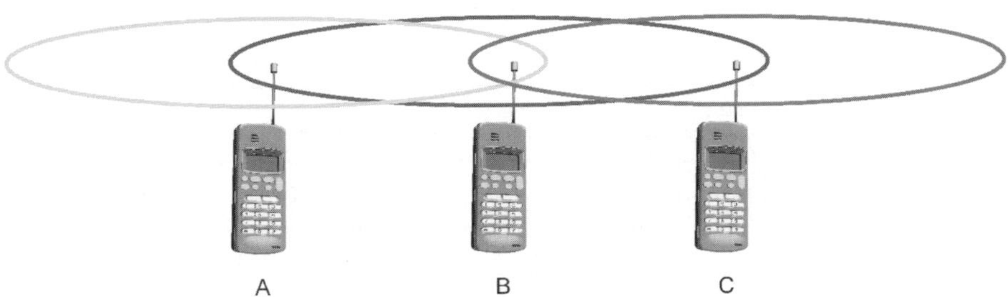


*Figure 3.13.* The hidden terminal problem: Node A is cannot hear node C, and vice versa. Therefore, simultaneous transmissions destined to node B by node A and node C will result in collisions.

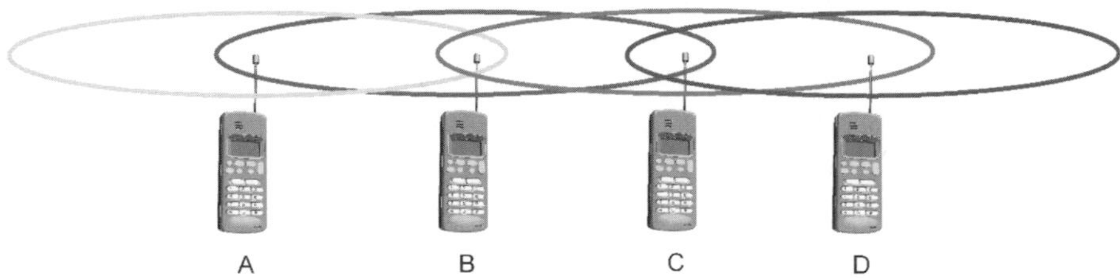


*Figure 3.14.* The exposed terminal problem. Node C is transmitting to destination D. Since the channel is busy due to node C's transmission, node B cannot transmit. However, node B's transmission for node A will not interfere with node C's transmission to node D. Thus, by preventing node B's transmission, bandwidth is wasted due to the underutilization of the channel.

any node that hears an ongoing transmission transmits a busy tone; any node that hears a busy tone does not initiate a transmission. Thus, all nodes in the two-hop neighborhood of a source node are silenced for the duration of the packet transmission. BTMA requires each node to have a multi-band radio or multiple radios (one for data and one for signaling). Furthermore, the solution for hidden nodes used in BTMA creates another problem, which is known as the exposed terminal problem (see Figure 3.14).

Handshaking is another solution for avoiding the hidden node problem, and it is the most popular method used in distributed MAC protocols, such as MACA [75], MACAW [33], and IEEE 802.11 [8]. The basic design principle is that the nodes around the transmitter and the receiver should be silenced during data transmission via pre-transmission messages (*i.e.*, Request-To-Send (RTS) and Clear-To-Send (CTS)) and post-transmission MAC level confirmation messages (*i.e.*, Acknowledgments (ACKs)). Handshaking is an efficient method to reduce collisions provided that the data packets are much larger than the control packets.

Figure 3.15 illustrates the four-way handshaking as it is implemented in the Distributed Coordination Function (DCF) of IEEE 802.11. When a node has data to transmit, it picks a random wait period (defer time). This wait period

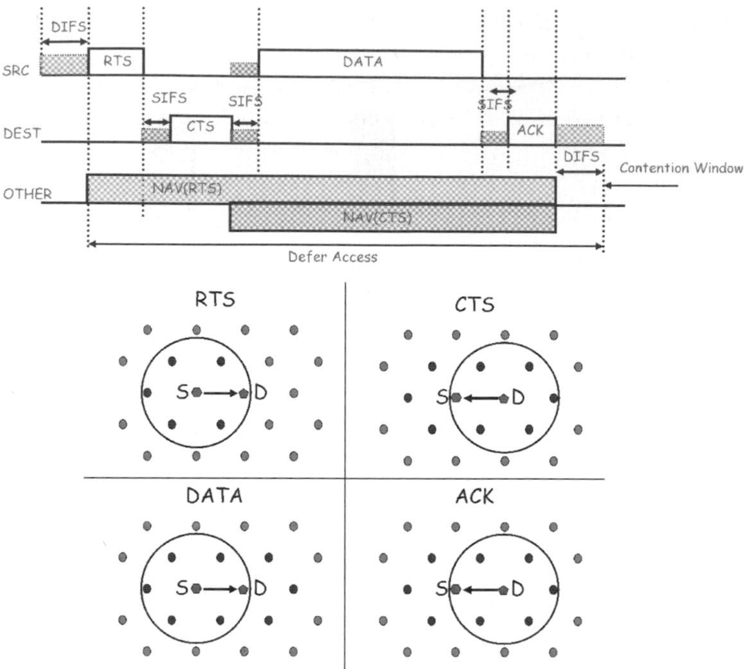


*Figure 3.15.* Illustration of IEEE 802.11 DCF four-way handshaking.

is decremented when the channel is idle at each time slot (*i.e.*, time is divided into slots). Upon the expiration of the defer timer, the node tries to acquire the channel by sending an RTS packet. This portion of the channel access is equivalent to $p$-persistent CSMA. The receiving node responds with a CTS packet indicating it is ready to receive data. Both the RTS and CTS packets contain the total duration of the transmission (*i.e.*, the overall time interval needed to transmit the data frame and the related ACK). Once an RTS or CTS is heard by the nodes in the one-hop neighborhood of the transmitter or receiver, they stop their defer timers and set their Network Allocation Vector (NAV) to the duration of the transmission. Thus, they cannot initiate an RTS nor can they respond to an RTS with a CTS. Upon the expiration of NAV, silenced nodes in the one-hop neighborhood of the sender and destination restart the countdown of their defer timers from the value at which they were stopped. This is called virtual carrier sensing.

Once an RTS-CTS exchange is successful, the sender than transmits the data packet. If the data packet is received successfully (*i.e.*, no collision or bit-errors), the destination node responds with an ACK. If an ACK is not received, the packet is assumed to be lost. If the handshaking fails at any point, then the transmitter starts over again.

The random defer time is picked form a uniform distribution with a minimum of zero and a maximum of the current value of the defer period. At each failure the defer period is doubled (up to a predefined maximum), and with each successful completion of a complete handshaking cycle the defer period is decremented linearly (down to a predetermined minimum, which is also the default value of the defer period). This contention resolution method is called Binary Exponential Backoff (BEB) with exponential increase and linear decrease. The defer period is the equivalent of the probability of transmission (*i.e.*, $p$) in $p$-persistent CSMA.

To ensure the proper operation of the handshaking cycle and enhance the protocol's robustness against various factors, such as dynamic propagation and interference characteristics, mobility, and packet errors, different waiting intervals are specified. A node needs to sense the channel idle for a Distributed Inter-Frame Space (DIFS) interval before making an RTS attempt and a Short Inter-Frame Space (SIFS) interval before sending any of the CTS, Data, or ACK packets. Since the SIFS interval is shorter than the DIFS interval, the station sending any one of the CTS, Data, or ACK packets attempts transmission before a station attempting to send an RTS packet and hence the handshaking interval is not interrupted prematurely.

IEEE 802.11 uses $p$-persistent CSMA in broadcasting. Since in broadcasting it is not possible to use handshaking, none of the advantageous features in unicasting, like BEB and NAV, can be utilized. Unlike unicasting, where the defer period is adjusted adaptively by using the BEB algorithm with the feedback information obtained from the success or failure of the handshaking cycle, in broadcasting it is not possible to adjust the defer period due to the lack of reliable channel feedback; hence, the defer period is constant and independent of the traffic conditions (*i.e.*, the default minimum defer period).

The Seedex protocol [76] avoids collisions by creating a distributed transmission schedule through exchange of the transmission schedules in a two-hop neighborhood, which is actually the whole network from the point of the view of the MAC layer. Each node creates its transmission schedule by using a Bernoulli process with parameter $p$. The information to be propagated is very compact, thus the overhead for the maintenance of the distributed collision-free transmission schedule is low.

Table 3.1 shows the media access used in several wireless systems. Systems that must guarantee a certain quality of service (QoS) to the user, such as cellular systems, typically use fixed-assignment multiple access techniques. In contrast, systems that make no guarantees about timely delivery of data, such as wireless data networks, often use random-access techniques. Several systems combine media access technologies, such as using TDMA with a CDMA protocol (*e.g.*, PCS-2000 [69]), assigning slots to users within a given frequency-hoping spread-spectrum (FHSS) protocol (*e.g.*, Bluetooth [3]), or using either TDMA

*Table 3.1.* Media access used in different wireless systems.

| System | Media access |
|---|---|
| AMPS | FDMA |
| PACS | TDMA |
| PCS-2000 | CDMA/TDMA |
| GSM | TDMA |
| Lucent WaveLAN | IEEE 802.11 |
| Bluetooth | FHSS/TDMA |
| HomeRF | FHSS/TDMA/CSMA |

or CSMA within a FHSS protocol (*e.g.*, HomeRF, where TDMA is used for real-time data delivery and CSMA is used for asynchronous delivery) [77].

# Chapter 4

# ROUTING

In multi-hop networks, source and destination nodes can be separated by multiple hops, and thus packets from the source to the destination need to be forwarded by multiple nodes. This forwarding process is known as routing. If there is one destination, then this type of data flow is called unicast routing. If there are multiple destinations or all the nodes in the network are destined by the source, then these types of flows are known as multicast routing and broadcast routing, respectively (see Figure 2.7).

## 4.1 Unicast Routing

Although it is possible to classify unicast routing protocols into many categories based on different criteria, categorization based on route discovery (*i.e.*, proactive routing protocols and reactive routing protocols) has found wide acceptance.

In proactive routing protocols each node keeps a routing table to all the other nodes in the network so that when a packet needs to be forwarded, the route is already known and can be immediately used. Each entry in the routing table contains the path (*i.e.*, node IDs in the path in ordered form). The routing table is updated periodically through control packet exchanges. Proactive routing protocols have the advantage that a node experiences minimal delay whenever a route is needed, as an already available route is immediately selected from the routing table. However, proactive routing protocols are not scalable, which means that for large networks the algorithm is not feasible. This is because maintenance of a complete routing table by each node consumes a substantial portion of the available bandwidth for relatively small networks, but in larger networks even using all of the bandwidth is not enough for routing table maintenance.

The Destination-Sequenced Distance Vector (DSDV) [78] protocol is a proactive routing protocol. In DSDV each node periodically broadcasts routing updates. Each node maintains a routing table for all possible destinations within the network. Each entry in the routing table is marked with a sequence number assigned by the destination node. The sequence numbers help identify the obsolete routes from the updated ones, which alleviates the formation of routing loops. Routing table updates are periodically propagated throughout the network to maintain consistency in the routing tables.

Reactive routing protocols, on the other hand, employ a Just-In-Time (JIT) approach, where nodes only discover routes to destinations on demand (*i.e.*, a node does not need a route to a destination until that destination is to be the sink of the data packets sent by the node). Reactive protocols often consume much less bandwidth than proactive protocols, but the delay to determine a route can be significantly higher.

Dynamic Source Routing (DSR) is a reactive routing protocol [62]. In DSR, each node keeps a route cache containing full paths to known destinations. If a node has no route to a destination, it broadcasts a route request packet to its neighbors. Any node receiving the route request packet that does not have a route to the destination appends its own ID to the packet and rebroadcasts the packet. If a node receiving the route request packet has a route to the destination, the node replies to the source with a concatenation of the path from the node to itself and the path from itself to the destination. If the node already has a route to the source, the route reply packet will be sent over that route. Otherwise, the route reply packet can be sent over the reversed source to node path, or piggybacked in the node's route request packet for the source. If an intermediate node discovers a broken link in an active route, then it sends a route error packet to the source, which may re-initiate route discovery if an alternate route is not available.

## 4.2 Multicast Routing

Both broadcasting and unicasting are special forms of a more general networking operation, which is multicasting. In multicasting, one or more source nodes convey information to the members of a multicast group, possibly through the use of non-multicast group member nodes within the network.

Multicast routing of voice traffic within a mobile ad hoc network has many applications, especially in military communications. For example, members of a medical or engineering unit within a larger formation of soldiers need a multicasting platform for their group communication needs. Furthermore, it is not possible to restrict the communication platform to a single-hop networking framework. In many situations a platform restricted to single-hop communications will not be enough to fulfill the connectivity requirements of a mobile group. For example, some of the members of a multicast group will not be in reach of a source that is beyond their single-hop transmit/receive range due to extended distance, obstacles or interference. Thus, the need for multi-hop voice multicasting is obvious within a wireless mobile ad hoc networking framework.

The first objective of a group communication protocol, in general, and a multicast protocol, in particular, is to convey packets from a source to the members of a multicast group with an acceptable Quality of Service (QoS) [79]. Actually, flooding (see Section 4.3.2), which is the simplest group communication algorithm, is good enough to achieve high Packet Delivery Ratio (PDR), provided that the data traffic and/or node density is not very high so that the network is not congested. However, flooding generally is not preferred as a multicast routing protocol due to its excessive use of the available bandwidth. In other words inefficiency of the spatial reuse of flooding prevents its use as an effective multicast routing protocol.

Thus, the second objective of a group communication protocol is to maximize the spatial reuse efficiency [20], which is directly related with the number of retransmissions required to deliver each generated data packet to all members of a multicast group with a high enough PDR. The third objective of a multicast protocol is to minimize the energy dissipation of the network. Minimizing the energy dissipation is crucial to keep the mobile users, equipped with lightweight battery-operated radios, connected to the network [38].

There are many multicast routing protocols designed for mobile ad hoc networks, which can be categorized into two broad categories: (*i*) tree-based approaches and (*ii*) mesh-based approaches. Tree-based approaches create trees originating at the source and terminating at multicast group members with an objective of minimizing a cost function. For example, shortest path tree algorithms [80] create trees originating at the source with an objective of minimizing the distance between the source and every destination in the multicast group individually. Minimum cost tree algorithms [81] minimize the cost function associated with the global multicast tree as a whole to create multicast trees. Constrained tree algorithms [82] extend the definition of the cost function from number of hops to other metrics, such as delay.

A multicast protocol for ad hoc wireless networks (AMRIS) [83] constructs a shared delivery tree rooted at one of the nodes with IDs increasing as they radiate from the source. Local route recovery is made possible due to this property of IDs, hence reducing the route discovery time and also confining route recovery overhead to the proximity of the link failure.

Mesh-based multicasting is better suited to highly dynamic topologies, simply due to the redundancy associated with this approach. In mesh-based approaches there is more than one path between the source and multicast group members; thus, even if one of the paths is broken due to mobility the other paths are available. On Demand Multicast Routing Protocol (ODMRP) [84] is a mesh-based scheme using a forwarding group concept, where only a subset of nodes forwards the multicast packets via scoped flooding. Instead of using a tree, ODMRP utilizes a mesh structure, which is redundant and robust, to compensate for the frequent route failures and trades-off bandwidth for stability, which comes with redundancy. ODMRP employs on-demand routing techniques to avoid channel overhead and improve scalability. Broadcasting is an important special case of multicasting, where the multicast group consists of all the nodes in the network; thus, we will discuss network-wide broadcasting in the following section.

## 4.3 Broadcasting Routing

Real-time data broadcasting is an important service in mobile ad hoc networks. In many applications, real-time data need to be broadcast throughout the entire network in a multi-hop fashion. For example, the leader of a search and rescue team may need to communicate with all members of the team connected to the network, or the soldiers in a battlefield mission may need to utilize the surveillance information of the region that they are operating within, broadcast by an observer located at a strategic position. Network-wide broadcasting algorithms can be classified into three main categories: (*i*) fully coordinated, (*i*)non-coordinated, and (*i*) partially coordinated.

### 4.3.1 Fully Coordinated Broadcasting Algorithms

The goal of a fully coordinated algorithm is to create a Minimally Connected Dominating Set (MCDS), which is the smallest set of rebroadcasting nodes such that the set of nodes are connected and all non-set nodes are within transmit range of at least one member of the MCDS [20]. An MCDS is the global optimal broadcasting scheme. However, implementation of such an algorithm is not practical, even with the assumption of global knowledge, due to the NP-hardness of the problem.

### 4.3.2 Non-Coordinated Broadcasting Algorithms

Flooding is an example of a non-coordinated broadcast algorithm, where nodes rebroadcast without any coordination [85, 86]. In order to avoid excessive collisions, nodes retransmit with a Random Assessment Delay (RAD), which is uniformly distributed in [0, $T_{RAD}$]. However, especially in dense networks, flooding is highly ineffective due to the excessive redundant rebroadcasts.

Gossiping is another example of a stateless (non-coordinated) broadcast algorithm [15, 52], where nodes rebroadcast with a predetermined probability $p_{GSP}$ in conjunction with RAD. However, regardless of $p_{GSP}$, source nodes always transmit. In gossiping the overall rebroadcast probability is an exponentially decreasing function of hop count. Thus, nodes close to the source receive many redundant versions of the same broadcast packet while farther nodes may not receive the packet at all.

### 4.3.3 Partially Coordinated Broadcasting Algorithms

Partially Coordinated Broadcast (PCB) algorithms can further be classified into two sub-categories: passive PCB and active PCB. The design principle of passive PCB algorithms is analogous to the binary exponential backoff scheme of IEEE 802.11, where the backoff window size is adjusted adaptively by passive listening of the medium. Counter-based broadcasting (CBB) and Distance-based broadcasting (DBB) are two examples of partially coordinated broadcast algorithms [15, 20, 87]. In CBB, a node that receives a packet randomly chooses its RAD and starts to count the number of receptions of the same packet until its broadcast timer expires. If the number of receptions of the packet is lower than the predetermined maximum counter value, $N_{CBB}$, then the packet is transmitted, otherwise it is dropped. DBB is a distance based scheme, where the nodes calculate the distance to a transmitting node based on received power strength. In DBB, each node picks a RAD upon reception of a previously unheard packet and starts to record the distance of the nodes that retransmit the same packet. Upon the expiration of RAD, if the closest transmission of the packet to be transmitted is higher than the minimum distance, $D_{DBB}$, than the packet is transmitted, otherwise it is dropped.

Active PCB algorithms can be considered as approximate limited scope MCDS's based on one-hop or two-hop neighborhood and/or topology information. In the algorithms proposed in [14, 16] a node makes a local decision to rebroadcast a packet if its set of neighbors is not the same as that of upstream nodes, where the neighbor information is exchanged through periodic hello messages. Algorithms proposed in [17, 18] are also based on two-hop neighborhood information exchange, but the decision to rebroadcast is made directly or indirectly by the upstream nodes. Broadcasting through clustering [12] also falls in the category of active PCB algorithms.

## 4.4 Hierarchically Organized Networks

Apart from the general classification of routing algorithms in flat networks, hierarchically organized networks also create options for unicasting, multicasting, and broadcasting. There are many studies on hierarchically organized networks [88–92, 10, 93, 94]. The first step in such networks is to create a hierarchy among the nodes through clustering.

### 4.4.1 Clustering

Achieving the goals of QoS and energy efficiency in a multi-hop network necessitates coordination between the nodes, so that they avoid wasting system resources. While these goals can be met using centralized control (*i.e.*, a unique centralized controller), this is not scalable due to the high overhead to monitor and convey the control information throughout the network. In fact, for large networks most of the bandwidth will be used for control traffic (*i.e.*, topology maintenance, channel access overhead, *etc.*). Furthermore, if the network size becomes very large the available bandwidth will not be enough to convey the enormous volume of the control traffic needed to maintain the network. Although completely decentralized coordination (*i.e.*, each and every node in the network is an independent controller) is realizable and practical when compared to centralized control, due to the lack of reliable coordination, its performance becomes unstable as the network load increases. The performance degradation of ALOHA with increasing node density and data traffic exemplifies the shortcomings of the decentralized approach. Thus, the quest for a solution where a sufficient level of coordination is supported within a mobile ad hoc network without harming the scalability of the network has resulted in clustering.

In a clustered network some of the nodes are elected as clusterheads and they act like local controllers for the nodes around them. A clusterhead and the nodes controlled by it constitute a cluster. Note that there may be multiple clusterheads for each cluster, and an ordinary node can be a member of multiple clusters. Nevertheless, we will assume that there is a single clusterhead for each cluster and any ordinary node is a member of a single cluster at a time for the sake of clarity at this point. The number of clusterheads are directly proportional with the network area and/or the number of nodes in the network, and the number of nodes in each cluster is independent of the network size. Thus, the volume of control traffic for topology control and channel access is independent of the network size in clustering. Hence, clustering provides a level of coordination higher than completely decentralized control and lower than centralized control while avoiding the scalability bottleneck of centralized control. While clustering enables both scalability and a sufficient level of coordination in the network, it also introduces the issues of clusterhead selection and clustering structure maintenance, which necessitate dedication of a portion of the resources (*e.g.*, bandwidth) for these operations. There is a wealth of clustering algorithms in the literature. We present some of the representative clustering algorithms and architectures in the following section.

### 4.4.2 Clustering Algorithms

One of the earliest studies on clustering is [89]. This clustering algorithm is based on the one-hop and two-hop neighbors of a node. In the highest-ID clustering algorithm each cluster is assigned a unique frequency band, thus inter-cluster interference is avoided by Frequency Division Multiple Access (FDMA). Clusters are linked through relay nodes or a direct link between the clusterheads. The network is clustered periodically, one cluster at a time. Channel access during clustering is determined by a fixed Time Division Multiple Access (TDMA) scheme, where slot assignment is based on node-ID (*i.e.*, node 1 transmits in slot 1, node $N$ transmits in slot $N$ in an $N$-node network). Each node transmits a list of the nodes it can hear directly (*i.e.*, its one-hop neighbors) in its reserved slot. At the end of $N$ slots, all the nodes have a complete list of their two-hop neighbors. None of the nodes have the complete connectivity matrix, but they have partial versions of it, which are consistent with the global connectivity matrix. The highest-ID clustering is a distributed implementation of the following centralized procedure. The highest-ID node (*e.g.*, node $N$) declares itself as a clusterhead. If there are any nodes left beyond the reach of node $N$ (*i.e.*, beyond its transmit range), then node $N-1$ becomes a candidate clusterhead, tentatively. If node $N-1$ covers any node that is not covered by node $N$, it becomes a clusterhead. Otherwise node $N-2$ becomes a candidate clusterhead and the procedure continues until all the nodes are within the range of at least one clusterhead.

The Near Term Digital Radio (NTDR) [95] uses a clustering approach with a two-tier hierarchical routing algorithm. Nodes form local clusters, and intra-cluster data are sent directly from one node to the next, whereas inter-cluster data are routed through the clusterhead nodes. This design allows for increased system capacity by reducing interference. In NTDR networks, the clusterhead nodes change as nodes move in order to keep the network fully connected. This protocol, designed to be used for a wireless data network, enables point-to-point connectivity.

A lowest-ID clustering technique is presented in [10]. In this technique, during network initialization, nodes decide on their status as a cluster leader or an ordinary node based on their IDs (see Figure 4.1). The clustering algorithm assumes all the nodes are aware of the IDs of their one-hop neighbors. If a node is the lowest-ID node among its neighbors, it becomes the cluster leader. An ordinary node that is in the transmission range of multiple cluster leaders joins the cluster with the lowest cluster ID, which is the same as that of the cluster leader. Inter-cluster communication flows through the relay nodes, which are ordinary nodes that are in the transmission range of multiple clusters. Transmission range and node density are the primary factors that determine connectivity and the number of repeaters needed. Therefore, the transmission range should be selected carefully to keep the network connected. In the case of mobile nodes, nodes can move out of the cluster leader's transmission range, and the number of hops between the nodes in a cluster may exceed two hops. In this case, the cluster should be reconfigured. The reconfiguration of the cluster is based on the highest connectivity. The node with the highest connectivity (*i.e.*, the highest number of one-hop neighbors) becomes the cluster leader and all the nodes in its one-hop neighborhood, which are not in the transmission range of another higher connectivity cluster leader, will join this cluster. During

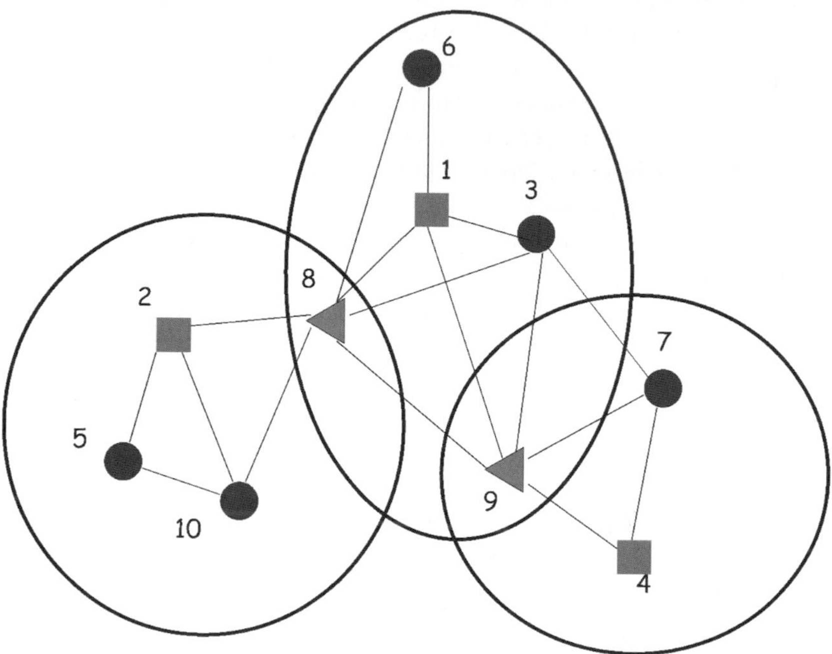


*Figure 4.1.* Illustration of the lowest-ID clustering algorithm. Squares, triangles, and disks represent clusterheads, gateways, and ordinary nodes, respectively.

network initialization, all the nodes in the network are communicating through a common Code Division Multiple Access (CDMA) code, but after the network is partitioned, clusters choose different CDMA codes for intra-cluster communication to avoid interference between the clusters. CDMA code selection is done by negotiation between the neighbor clusters. Intra-cluster medium access for an $N$-node cluster is through a fixed TDMA schedule organized into frames, which have $N + 1$ slots for packet transmission. Each node has a single slot for transmission in each frame. The last slot serves as the temporary slot for a new node until the fixed TDMA frame is recomputed. The repeater nodes listen to the CDMA codes of the neighboring clusters randomly, and thus there is a non-zero probability that a repeater catches a packet intended for it.

Power control can be used to dynamically adjust the size of clusters [96]. If open-loop power control is used, the clusterhead node sends out a beacon, and nodes that hear the beacon join the cluster. If there are too many nodes in the cluster, the clusterhead can reduce the beacon signal strength so fewer nodes will hear it. On the other hand, if the cluster is too small, the clusterhead can increase its beacon signal strength to increase cluster membership. New clusters may be formed when a clusterhead decreases its membership size, and clusters may be merged when a clusterhead increases its membership size in order to keep the network fully connected.

A clustering approach based on lowest-ID and node mobility patterns is described in [88]. Each node is assumed to be equipped with GPS. Clusters are formed around the nodes that are more stationary in a neighborhood. For example, in a mobility model where nodes are moving as a group, the node that has the closest motion pattern to the average group motion pattern is selected as the clusterhead. It is assumed that each node is aware of the mobility of all the nodes in their one-hop neighborhood.

McDonald *et al.* develop a clustering algorithm that enables good routing (high probability of path availability) while supporting node mobility and stability [97]. Their $(\alpha, t)$ cluster algorithm creates clusters of nodes where the probability of path availability is bounded over time. This allows the clustering algorithm to adapt to node mobility, creating more optimal routing under low node mobility.

The distributed and mobility adaptive clustering (DMAC) algorithm is introduced in [90, 91]. This is a flexible algorithm in the sense that the criterion to become a clusterhead is not specific and is defined by a generic weight function, which can be application driven. For example, node speed can be taken as a weight function, which results in a lower number of clusterhead changes when compared to a lowest-ID algorithm. Transmit power level can also be used as the weight function. In energy aware protocols, energy level in the nodes can be used as a weight function. Cluster switching and clusterhead resignation or initiation are not decided by sharp limits; instead, a variable threshold is incorporated when comparing the weights, which helps avoid frequent changes in the clusters. The DMAC algorithm is compared to the lowest-ID clustering algorithm [10], and it is reported that DMAC outperforms the lowest-ID algorithm as much as 85% in terms of clustering overhead.

In [98], a hierarchical multi-hop network architecture, which partitions the network into clusters organized around special nodes (switches), is proposed. The network organization is hierarchical with multiple levels. Clustering in this study is different from the other studies [88–92, 10], where the clusterheads are ordinary nodes.

A hierarchical routing protocol using IEEE 802.11 as the MAC layer is presented in [93], where the network is partitioned into k-hop clusters and the clustering structure is shown to be stable due to the coordinated mobility of the nodes (*i.e.*, relative mobility of the nodes within the same cluster is negligibly small). Clusterheads are elected from the backbone nodes (BN), which are high power radios capable of traversing multiple hops of the ordinary nodes in a single-hop. BN's form a backbone network among themselves for routing.

A simulation based comparative evaluation of various clustering schemes is presented in [92]. The existing clustering algorithms are divided into five

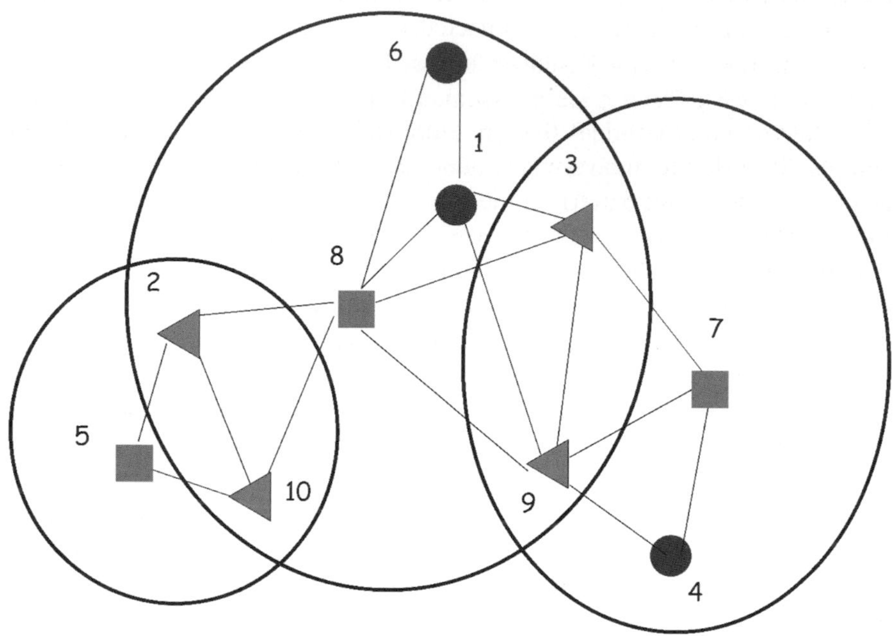


*Figure 4.2.* Illustration of the highest degree (connectivity) clustering algorithm. Squares, triangles, and disks represent clusterheads, gateways, and ordinary nodes, respectively.

categories. The first one is the highest connectivity algorithm (HC), which is based on cluster creation around the highest connectivity node (see Figure 4.2). The main drawback of this algorithm is the frequent clusterhead changes due to mobility.

The second algorithm is the lowest-ID clustering algorithm. In the lowest-ID algorithm, nodes are clustered around the lowest-ID nodes, which is reported to result in fewer clusterhead changes than the HC algorithm because the connectivity of a node changes frequently, which necessitates clusterhead switching.

The third algorithm is the Least Cluster Change (LCC) algorithm. Actually, this is not a stand alone algorithm but a cluster maintenance scheme that can be used in conjunction with the HC or lowest-ID algorithms. This scheme restricts the clusterhead changes to two cases, which are either a node gets disconnected from all the clusterheads or two clusterheads come into transmission range of each other. The LCC algorithm increases the stability of clusters when compared to the HC and Lowest-ID algorithms.

The fourth clustering algorithm is the distributed mobility adaptive clustering (DMAC) algorithm [90, 91]. As described previously, in DMAC, nodes are clustered around the highest weight node. The weight function is generic (*i.e.*, ID, connectivity, power, speed).

The fifth algorithm is the Weighted Highest Degree (WHD) clustering algorithm. In this scheme, when two clusterheads come into transmission range of each other, both clusters are decomposed and re-clustered, which results in a lower number of clusters but stability also degrades.

Distributed label clustering (DL) is proposed in [92], which chooses clusterheads according to a weight that maximizes the cluster size based on the sum of the degrees of the neighbors. The DL algorithm avoids making the leaf nodes clusterheads. Cluster maintenance is based on the LCC algorithm. The simulation results show that the LCC algorithm performs better in terms of fewer clusterhead changes and cluster switching. The authors used a random way point mobility model to simulate the effects of mobility for various speeds. The DL algorithm is shown to be the best clustering scheme in the majority of the simulations. The price paid for the increase in performance is increased information exchange between the neighbors. However, since there are fewer cluster changes in the DL scheme, the number of packets for cluster maintenance is also reduced. The important conclusion from this work is that the LCC scheme gives more stable clustering results than the other algorithms that do not have cluster maintenance schemes but instead employ re-clustering schemes when the topology changes due to mobility.

Chapter 5

# ENERGY EFFICIENCY AND QOS

This chapter provides a bridge between the previous background chapters and the later research chapters of this book. In this chapter we present the necessary background on energy efficiency and Quality of Service (QoS) in mobile ad hoc networks.

## 5.1 Energy Efficiency

Mobile radios rely on batteries, which are limited sources of energy. Thus, optimizing the energy dissipation of both the individual radios and the total network is one of the major considerations in designing algorithms for Mobile Ad Hoc Networks (MANETs). Experimental results have revealed that 50% of the overall energy dissipation of handheld devices is due to networking related activities [99, 100].

### 5.1.1 Power Consumption Characteristics

Power consumption of a wireless radio depends on the operation mode. Operation modes of a radio can be categorized into the following: (*i*) transmit mode, (*ii*) receive mode, (*iii*) idle mode, and (*iv*) sleep mode [101]. The actual power consumption values depend strongly on the hardware implementation of the radio. Table 5.1 presents the power consumption of several Network Interface Cards (NIC) in transmit, receive, idle, and sleep modes.

In order to obtain some insight into the differences between the power consumptions in different operation modes we will concentrate on the Aironet PC4800 PCMCIA NIC [102]. Figure 5.1 shows the power consumption values for the transmit, receive, idle, and sleep modes for this card, and the schematic

*Table 5.1.* Power consumption values of some IEEE 802.11 (2.4 GHz) wireless cards in different operation modes (mW).

| Radio | Transmit | Receive | Idle | Sleep |
|---|---|---|---|---|
| Aironet PC4800 [102] | 2500 | 900 | 110 | 20 |
| Aironet 350 PCI [103] | 1870 | 1620 | 1440 | 910 |
| Lucent Bronze [104] | 1300 | 970 | 840 | 66 |
| Lucent WaveLAN [105] | 1400 | 1200 | 1000 | 150 |
| Cabletron Roamabout [106] | 1400 | 1000 | 830 | 130 |

of the card is shown in Figure 5.2. Table 5.2 presents the power consumption values of each individual component of the card at different operating modes.

Transmit power is used for packet transmissions. Most of the energy is dissipated on the power amplifier (*i.e.*, 64% of the total transmit power is for the power amplifier in the transmit mode). Some of this power is used by the circuitry itself, while the rest is radiated into the channel to compensate for the path loss (see Section 2.2.1). Digital processing consumes only a tiny fraction of the total power in transmit mode (*i.e.*, the MAC and baseband processors consume only 6.4% of the total transmit power).

Receive power is used for packet receptions. The Intermediate Frequency (IF) modem consumes most of the power in reception (*i.e.*, 55%), and the power amplifier is turned off. In receive mode digital processing consumes approximately one fourth of the total power.

Some radios are designed in a more elegant way than others (*e.g.*, the Aironet PC4800 PCMCIA NIC), and they can distinguish between a clear channel and a busy channel. If the channel is sensed busy (*i.e.*, a transmission is going on) by the Low Noise Amplifier (LNA), then the rest of the receive circuitry (*e.g.*, RF/IF converter, dual frequency synthesizer, *etc.*) actively processes the data coming up from the LNA. However, if the channel is sensed clear by the LNA, then the rest of the circuitry is kept in a low energy state without actually processing any data coming up from the LNA. This mode of operation is known as idle mode. In the idle mode two thirds of the power consumption is due to digital processing. In sleep mode a node is not able to receive or transmit.

Sleep mode power is consumed by electronic circuitry to keep the radio in a low power state that can return back to active mode in reasonable time, when required. Furthermore, both the power amplifier and the LNA are completely turned off during the sleep mode.

The power consumption of the radio could be lowered by pushing the circuitry into deeper sleep modes. However, the deeper the sleep mode the higher the recovery time of the radio. Hence, even in the sleep mode some of the

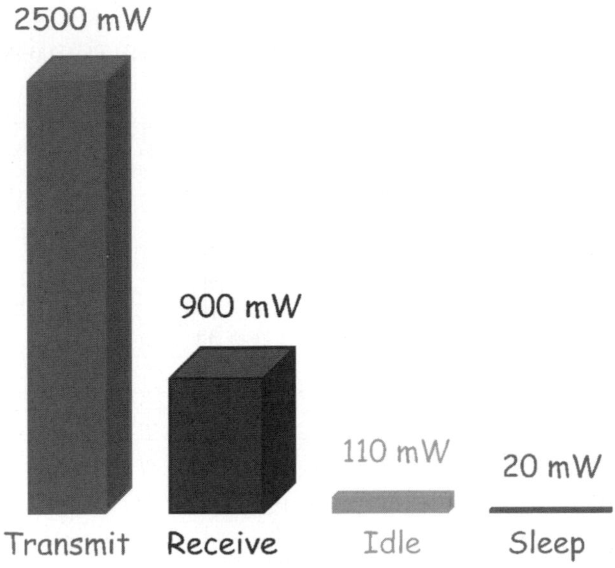


*Figure 5.1.* Aironet PC4800 PCMCIA Network Interface Card power consumption in transmit (2500 mW), receive (900 mW), idle (110 mW), and sleep (20 mW) modes.

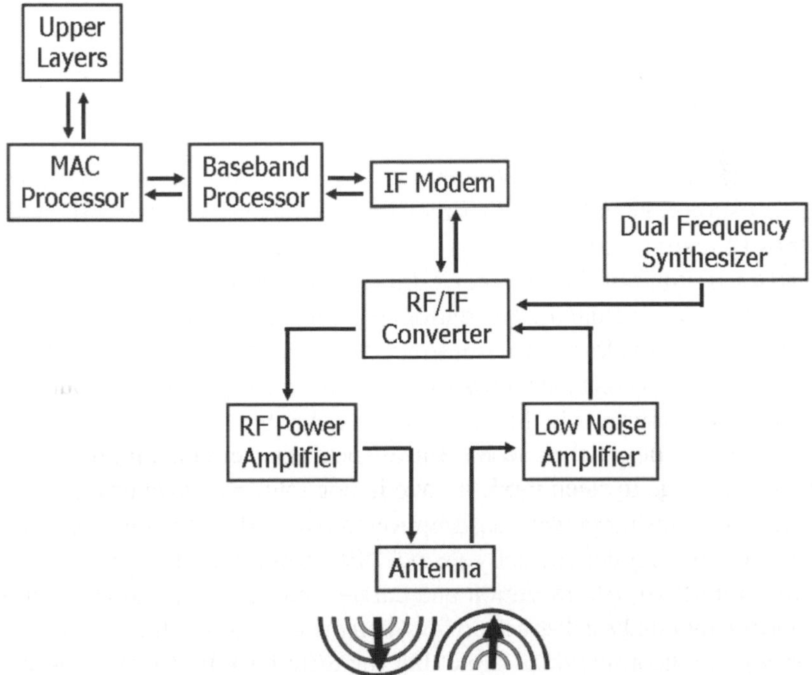


*Figure 5.2.* Schematic of Aironet PC4800 PCMCIA Network Interface Card.

*Table 5.2.* Power consumption of the components of the Aironet PC4800 PCMCIA Network Interface Card in mW.

| Circuit Component | Transmit | Receive | Idle | Sleep |
|---|---|---|---|---|
| MAC Processor | 125 | 125 | 40 | 5 |
| Baseband Processor | 33 | 100 | 23 | 2 |
| Intermediate Freq. Modem | 400 | 500 | 10 | 10 |
| Dual Freq. Synthesizer | 40 | 40 | 0.075 | 0.075 |
| RF/IF Converter | 300 | 100 | 0.05 | 0.05 |
| Low Noise Amplifier | off | 35 | 35 | off |
| RF Power Amplifier | 1600 | off | off | off |
| Max. Total Power | $\sim$ 2500 | $\sim$ 900 | $\sim$ 110 | $\sim$ 20 |

*Table 5.3.* Lucent WaveLAN IEEE Turbo 11 Mbps PC card sleep characteristics.

| Mode | Power Consumption | Recovery Time |
|---|---|---|
| Sleep Mode 1 | 950 mW | 1 $\mu$ s |
| Sleep Mode 2 | 350 mW | 25 $\mu$ s |
| Sleep Mode 3 | 300 mW | 2 ms |
| Sleep Mode 4 | 150 mW | 5 ms |

components of a radio are not completely turned off. Table 5.3 shows the power consumption and recovery times for the Lucent WaveLAN IEEE Turbo 11 Mbps PC card [105].

Power consumption in the receive mode can further be classified into two functional classes: actual data reception and carrier sensing [107]. If the transmitter node is in the effective transmit/receive range, then with a high probability the incoming data is received without bit errors (see Figure 3.12). On the other hand, if the data source is in the carrier sense range but further than the effective transmit/receive range, then with a high probability, the incoming data has bit errors due to the low signal strength. We refer to this mode of power consumption as carrier sensing power consumption, as the radio can sense an on-going transmission but cannot correctly decode the signal. Yet the power consumption for both actual data reception and carrier sensing is the same because the hardware cannot make a distinction between these modes without actually fully processing the incoming data. Note that the difference between the idle mode and the carrier sense mode is that in the idle mode the energy in the channel of the carrier is indistinguishable from that of pure noise, however, in carrier

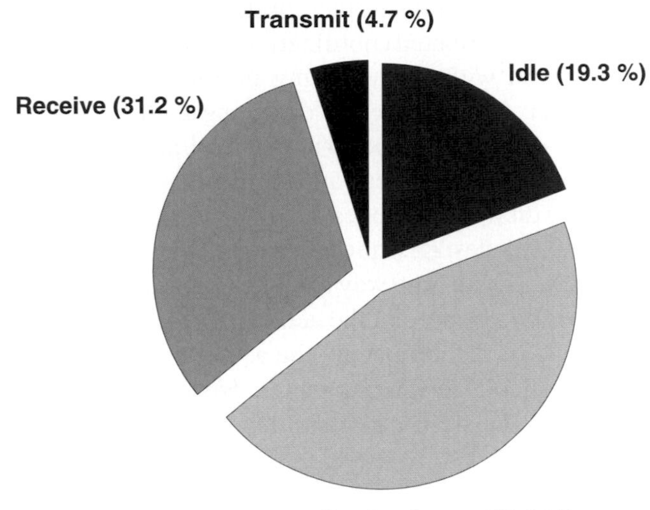


*Figure 5.3.* Energy dissipated on transmit, receive, idle, and carrier sense modes for flooding with IEEE 802.11 in an 800 m by 800 m network with 40 nodes.

sensing, although the received signal strength is low, it is distinguishably higher than that of the noise.

Throughout this book the energy model described in [38] is used. Power consumption in the transmit ($P_T$), receive ($P_R$), and idle ($P_I$) modes are 600 mW, 300 mW, and 100 mW, respectively. 318 mW of the transmit power is consumed by the circuitry and the rest (282 mW) is actually output from the antenna. In Chapter 6, Chapter 7, and Chapter 8 sleep mode power consumption is set to zero. In Chapter 9, Chapter 10, and Chapter 11 sleep mode power consumption is 10 mW.

To illustrate the energy dissipation characteristics of a simple network-wide broadcasting architecture (flooding using the IEEE 802.11 MAC), we present an example scenario. Figure 5.3 shows the relative amount of energy dissipation per node in the transmit, receive, carrier sense, and idle modes for an 800 m by 800 m area network with 40 nodes and a source sending data at 32 Kbps. Further details of this scenario can be found in Section 9.4.1. The largest component of energy dissipation is carrier sensing (44.9%), which is followed by receive energy dissipation (31.2%) and idle energy dissipation (19.3%). Transmit energy dissipation (4.7%) is the smallest component of the total energy dissipation. Since the underlying medium access control (MAC) protocol, which is IEEE 802.11, does not support a low-energy sleep mode in ad hoc (infrastructureless) mode for broadcasting, energy dissipated in the sleep mode is zero.

In general, energy-efficient distributed protocol design can be described as creating an appropriate distributed coordination scheme that minimizes a radio's total energy dissipation without sacrificing its functionality, by intelligently switching between the radio's different operating modes. Actually, there are only three modes that a radio can be switched to: transmit mode, active mode (receive, carrier sense and idle modes), and sleep mode. Although further classification of the energy dissipation modes of a radio is possible (*i.e.*, deep/shallow sleep modes, transient modes, *etc.*), the aforementioned classification is detailed enough in this context. There is no way to switch between receive, idle, and carrier sense modes: when a node is in the active mode, the actual mode (receive, idle or carrier sensing) is determined by the activities of the node's neighbors, which is not a controllable design parameter. Nevertheless, the ultimate goal is to keep the radio in the sleep mode as long as possible without sacrificing network performance.

In particular, energy efficiency in MANETs can be achieved by (*i*) avoiding unnecessary carrier sensing and minimizing the idle energy dissipation, (*ii*) avoiding overhearing irrelevant packets (*i.e.*, promiscuous listening), (*iii*) minimizing the transmit energy dissipation, by optimizing the transmit power and minimizing the number of retransmissions in broadcasting scenarios, and (*iv*) reducing the overhead (*i.e.*, bandwidth and energy used for anything other than optimal data transmission and reception) as much as possible without sacrificing the robustness and fault tolerance of the network [52, 108, 21, 22].

### 5.1.2 Idle Mode Energy Saving Techniques

Avoiding energy dissipation in the idle mode (idle and carrier sense energy) necessitates coordination through scheduling among the nodes [109], so that nodes avoid idle listening or carrier sensing. Many approaches have been proposed for minimizing the idle energy dissipation in single-hop wireless networks.

The IEEE 802.11 standard includes an energy saving mechanism when it is utilized in the infrastructure mode [8]. A mobile node that needs to save energy informs the base station of its entry to the energy saving mode, where it cannot receive data (*i.e.*, there is no way to communicate with this node until its sleep timer expires), and switches to the sleep mode. The base station buffers the packets from the network that are destined for the sleeping node. The base station periodically transmits a beacon packet that contains information about such buffered packets. When the sleeping node wakes up, it listens for the beacon from the base station, and upon hearing the beacon responds to the base station, which then forwards the packets that arrived during the sleep period. This energy saving method results in additional delays at the mobile nodes that may affect QoS. Furthermore, this approach is not directly applicable in multi-hop networks.

IEEE 802.15.3 is a dynamic TDMA based MAC protocol that is designed with multiple power management modes to support low power portable devices [110]. For example, if a non-controller device wants to be in power saving mode then it only listens to the beacon for an incoming message notification during its dedicated time slots. Additionally, it is possible for the devices in the network to enter into a very low power state by specifying a sleep period, which may span a long time.

The Energy Conserving Medium Access Control (EC-MAC) [109] protocol is designed for an infrastructure network with a single base station serving mobile nodes in its coverage area. In EC-MAC time is organized into cyclic time frames. Each frame starts with a Frame Synchronization Message (FSM) transmitted by the base station, which contains the synchronization information and the uplink transmission order for the subsequent reservation phase. During the request/update phase, each registered mobile node transmits new connection requests and the status of established queues according to the transmission order received in the FSM. A new user phase is used to register any new nodes that entered the coverage area by using S-ALOHA medium access. The base station transmits the schedule for the downlink and uplink transmissions in the schedule transmission slot. The rest of the time frame is used for downlink and uplink transmissions as specified in the transmission schedule. Due to its energy-efficient design through scheduling and cyclic time frame based channel access, the mobile nodes are able to maximize their sleep time. However, all of the aforementioned energy-efficient designs (*i.e.*, IEEE 802.11 infrastructure mode, IEEE 802.15.3, and EC-MAC) are confined to single-hop networks.

Several distributed MAC protocols have been developed with the goal of minimizing energy dissipation of the nodes. Sensor-MAC (SMAC) [111, 22] is an energy-efficient MAC protocol designed specifically for sensor networks that reduces idle listening by periodically shutting the radios off. All the nodes in the network synchronize through synchronization packet broadcasts in a master-slave fashion to match their non-sleep periods. Furthermore, overhearing is avoided by entering the sleep mode after receiving a Request-To-Send (RTS) and/or Clear-To-Send (CTS) packet until the Network Allocation Vector (NAV) timer expires, which is matched to the duration of the data packet. It is shown that in low traffic networks SMAC is much more energy-efficient than IEEE 802.11. Energy dissipation characteristics of SMAC are mainly determined by the sleep/active ratio, $R_{SMAC}$, and sleep/active cycle, $T_{SMAC}$.

### 5.1.3 Receive Mode Energy Saving Techniques

Especially in broadcasting, many redundant versions of the same packet are received by each node, which results in receive energy dissipation for no gain.

An efficient solution to this problem is information summarization prior to data transmission through a short information summarization (IS) packet that includes metadata summarizing the corresponding data packet transmission (*e.g.*, RTS/CTS packets of IEEE 802.11 in unicasting) [52]. A node that has already received a packet will be prevented from receiving redundant copies of the same packet, which are identified through corresponding IS packets, by entering the sleep mode.

Power Aware Multi-Access protocol with Signaling for ad hoc networks (PAMAS) [21] is an energy-efficient MAC protocol that is built on top of the Multiple Access Collision Avoidance (MACA) protocol [75]. In PAMAS nodes are equipped with two independent channels that are capable of transmitting and receiving without creating interference for each other, one for signaling and the other for data transmissions. Nodes avoid energy dissipation for overhearing packets destined for other nodes by entering the sleep mode. RTS/CTS packets are used to discriminate the data packets, thus, the metadata is the destination address of the unicast packets in this specific application. Due to the lack of RTS/CTS packets in broadcasting, it is not possible to employ PAMAS for broadcasting.

### 5.1.4 Transmit Mode Energy Saving Techniques

It has been shown that optimal network-wide broadcast scheduling for throughput or delay optimization in a multi-hop, mobile, packet radio network is NP-complete [13, 112]. Furthermore, it remains as an open question whether minimum transmit energy broadcast routing can be solved in polynomial time, despite the NP-hardness of its general graph version [113, 114, 54]. Minimum energy broadcasting is defined as finding a set consisting of relaying nodes and their respective transmission levels so that all nodes in the network receive a message sent by the source node, and the total transmit energy for this task is minimized [115]. Several approximation algorithms and their distributed versions for minimum energy broadcasting have been proposed [116–118]. In [119] three heuristic algorithms for the construction of the minimum energy broadcast tree computation are presented. Assumptions like complete knowledge of the node positions, a stationary network, an infinite number of frequencies or CDMA codes, no collisions and zero call blocking make these algorithms too restrictive to be used in an actual protocol. Furthermore, most of these algorithms tend to ignore the sources of energy dissipation other than transmit energy, such as energy dissipation for monitoring the network status and energy dissipated in receive, carrier sense, and idle modes.

The MiSer protocol minimizes the transmit energy consumption in IEEE 802.11a/h systems by transmit power control and physical rate adaptation [107].

The key idea is to create an optimal rate-power combination table to determine the most energy-efficient transmission strategy for each data frame.

By considering both transmit and receive energy dissipation, it has been shown that for a given energy and propagation model there is an optimum transmit radius, $D_{OP}$, beyond which single hop transmission is less energy-efficient than multi-hop transmissions [120, 121, 108]. Thus, the optimal broadcast strategy to minimize the transmit energy dissipation in a network consisting of constant transmit range radios is to use a multi-hop broadcasting scheme, where the transmit radius is chosen lower than $D_{OP}$. Furthermore, the total transmit energy dissipation increases with the number of retransmissions of a broadcast packet. Thus, reduction of the number of rebroadcasts results in higher energy savings.

## 5.2 Quality of Service

Quality of Service (QoS) is the performance level of a service offered by the network to the user [122]. A service can be characterized by a predefined set of measurable service requirements. QoS for streaming media throughout the network necessitates timely delivery of packets (bounded delay), high packet delivery ratio, and low jitter [123, 79]. Packet delay is directly related with the number of hops traversed by the packets and the congestion level of the network. In a highly congested network, packets are backlogged in the MAC layer before they can be transmitted, which increases the packet delay beyond acceptable limits. To ease congestion, packets that have exceeded the delay bound can be dropped rather than transmitting them to the destination, as they are no longer useful to the application. However, excessive packet drops decrease the packet delivery ratio, which is the other important aspect of QoS for streaming media. Packet delivery ratio is also decreased by collisions. Thus, there are two mechanisms that negatively affect the packet delivery ratio: packet drops and collisions.

### 5.2.1 QoS Metrics

The overall deterioration of QoS in voice communications can be expressed as the sum of individual factors, such as packet delay, packet loss, jitter, noise, and echo [123, 79]. Furthermore, the net effect of the distortion depends also on the codec specifications and the voice coding scheme utilized. For acceptable QoS in voice communications, the packet delivery ratio should be higher than a certain $\mathrm{PDR}_{MIN}$ in the absence of network delay, and the maximum network delay (excluding the delay contributions by various processing blocks, such as codec assembly and disassembly delays) should be less than a certain $\mathrm{Delay}_{MAX}$ in the absence of packet loss. The actual values of $\mathrm{PDR}_{MIN}$ and $\mathrm{Delay}_{MAX}$ depend on the voice codec. For example, $\mathrm{Delay}_{MAX}$ in lower

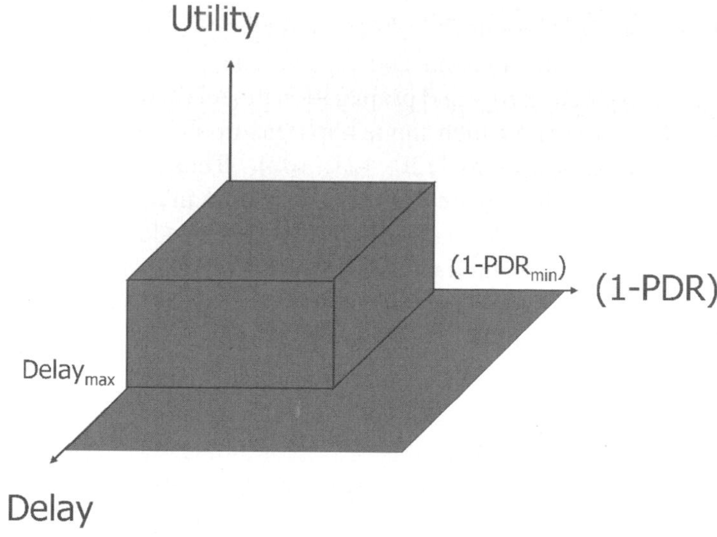


*Figure 5.4.* Delay-Packet Delivery Ratio (PDR) utility function.

bit rate voice coding is lower than Delay$_{MAX}$ in higher bit rate voice coding [123]. In Chapter 6, Chapter 7, and Chapter 8 we used a voice codec with Voice Activity Detection (VAD), which has a Delay$_{MAX}$ range of 30 ms to 50 ms [6, 7]. In Chapter 9, Chapter 10, and Chapter 11 we used a Constant Bit Rate (CBR) voice codec, which has a higher Delay$_{MAX}$ range (150 ms to 300 ms). Thus, the resulting utility function uses a hard constraint satisfaction scheme, where either the QoS is satisfied or not (see Figure 5.4) [124]. Although the utility function presented in Figure 5.4 is a rather simplified version of an actual utility function with higher dimensionality, we believe it satisfactorily captures the essence of the model for evaluating the QoS performance of network-wide voice broadcasting.

### 5.2.2 QoS Supporting Protocol Architectures

In single-hop and multi-hop broadcasting and multicasting scenarios, where acknowledged data delivery is not possible, QoS of the streaming media is determined primarily by the MAC layer. One solution to meet the delay, jitter, and packet delivery requirements for voice is to use periodic time-frame based medium access with automatic renewal of channel access, where the frame rate is matched to the periodic rate of the voice sources [6, 7]. This ensures that flows are uninterrupted, but it requires central control to coordinate channel access.

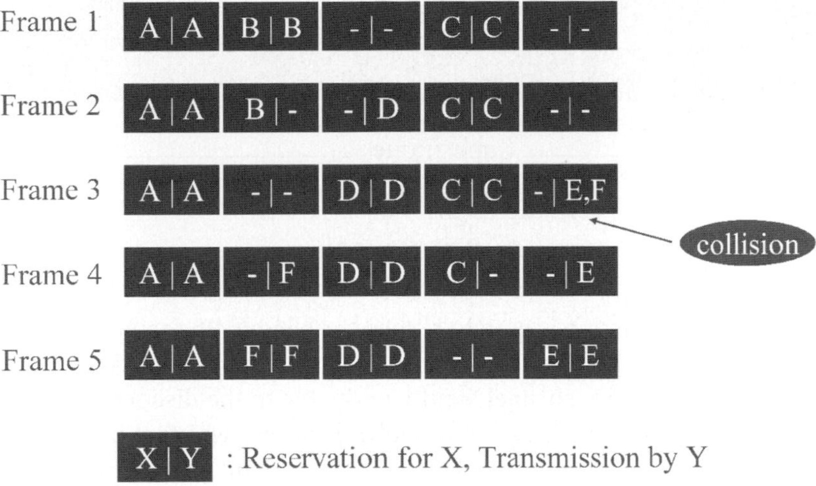


*Figure 5.5.* Illustration of R-ALOHA medium access control. Notation "X | Y" stands for "Reservation for X, Transmission by Y".

Continuation of data slot reservation for an uninterrupted sequence of data packets is the key feature of a real-time communication protocol that can provide QoS to multimedia applications, such as bounded delay and high packet delivery ratio for voice packets. In the rest of this section we will present the operation principles, advantages, and disadvantages of Reservation ALOHA (R-ALOHA) and Packet Reservation Multiple Access (PRMA), which are prominent examples of MAC protocols with QoS provisioning.

R-ALOHA, originally proposed for satellite communications, was the first protocol that employed the idea of slot reservation [4, 73, 32]. R-ALOHA is a combination of S-ALOHA and TDMA. In R-ALOHA, time is organized into frames, and frames are divided into slots. The frame structure of R-ALOHA is inherited from TDMA, which is illustrated in Figure 5.5. Successful data transmission in a slot automatically reserves the corresponding slot for the transmitting node in the next frame. By repeated use of that slot position, a node can transmit a long stream of data. Any unreserved slot is available for the next frame; nodes may contend for that slot using S-ALOHA. Thus in R-ALOHA, contention is on data slots and collisions corrupt (possibly long) data packets. All the nodes in the network should be on all the time in order to monitor the status of each slot. If there is a packet transmission, all the nodes receive it and discard it if it is not destined for them. Inherently it is not possible to save power with R-ALOHA. Fairness and prioritization are also not addressed by R-ALOHA.

| Contention Period | Contention-free Period |
|---|---|

← Superframe →

*Figure 5.6.* IEEE 802.15.3 superframe.

Voice activity detection is used to reduce the bandwidth required to transmit voice by not generating any packets during gaps in the speech. This increases the capacity of the radio channel significantly due to the discontinuous nature of speech. Voice activity detection in multiple access was first used in Packet Reservation Multiple Access (PRMA) [6, 7]. The main goal of PRMA, which is closely related with R-ALOHA, is to support real-time voice traffic and use the remaining bandwidth for asynchronous data transmissions. PRMA is distinguished from R-ALOHA by its response to network congestion and the use of voice activity detection. In PRMA, information packets from periodic sources, such as speech, are discarded if they remain in the node beyond a certain time limit. PRMA is designed to operate in a star topology, where the base station is in the center and the wireless nodes are around it. No direct communication is supported; even if the nodes are within communication range, they must communicate via the base station (*i.e.*, the same operation principle as Bluetooth). Energy efficiency and support for broadcast were also not among the design considerations of PRMA.

Stability is an important issue, which determines the system performance for R-ALOHA and PRMA [125, 126]. If the number of nodes contending for the same slot is too high, then none of the contending nodes can capture the data slot because of collisions. Therefore, both throughput and delay suffer severely. In order to sustain the system stability, the number of contending nodes and available data slots should be estimated and the system parameters should be updated accordingly [7, 126].

IEEE 802.15.3 [9] is a developing standard for single-hop networks to support applications with QoS requirements, such as video and voice. Time is organized into superframes consisting of a contention period, where contention for channel access and small bursty data are transmitted, and a contention-free period, where nodes transmit their data packets, based on the QoS requirements of the applications (see Figure 5.6).

Chapter 6

# SH-TRACE PROTOCOL ARCHITECTURE

## 6.1 Introduction

Many common applications require a peer-to-peer, single-hop, infrastructureless, reliable radio network architecture that enables real-time communication. Application areas of such networks include all kinds of group communications within a collection of mobile nodes that move according to a group mobility model, like the reference point group mobility model [63], without loosing full connectivity. In a single-hop radio network there are practically three independent entities above the physical layer: the Medium Access Control (MAC) layer, the transport layer and the application layer (*e.g.*, by definition, a routing layer is not necessary in a stand-alone fully-connected single-hop network).

In this chapter we present SH-TRACE [41, 44], a MAC protocol that combines different features of centralized and distributed MAC protocols to achieve high performance for peer-to-peer single-hop infrastructureless wireless networks (see Figure 6.1). SH-TRACE establishes one node in the network as a controller. This controller node coordinates channel access through a dynamically updated transmission schedule. To balance the coordinator energy load, SH-TRACE employs dynamic controller switching. Other features of SH-TRACE, such as information summarization, data stream continuation monitoring, multi-level controller backup, priority based channel access, and contention for channel access reinforce the energy efficiency, reliability, bounded delay, and maximized throughput of the network. Although SH-TRACE can be categorized as a MAC protocol, due to its cross-layer design it performs some of the functionalities of the other layers, such as data discrimination through information summarization, which is a function of the application layer.

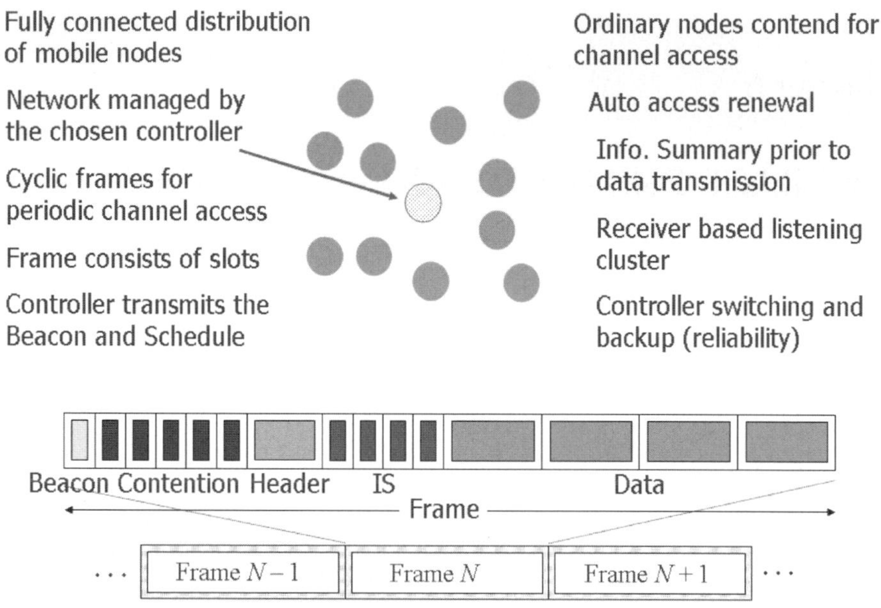
Fully connected distribution of mobile nodes
Network managed by the chosen controller
Cyclic frames for periodic channel access
Frame consists of slots
Controller transmits the Beacon and Schedule
Ordinary nodes contend for channel access
Auto access renewal
Info. Summary prior to data transmission
Receiver based listening cluster
Controller switching and backup (reliability)
Beacon Contention Header IS Data
Frame
...
Frame N − 1
Frame N
Frame N + 1
...

*Figure 6.1.* Overview of SH-TRACE operation.

SH-TRACE has been designed to be a very energy-efficient, reliable protocol to support real-time data broadcasting. Thus SH-TRACE is well suited to fulfill the tactical communication requirements of a small to medium size military group (*i.e.*, a squad) or a law enforcement group (*i.e.*, police officers pursuing a criminal or airport security personnel searching a group of passengers), where the members of the network may want to communicate simultaneously with each other. A group of researchers, students or tourists having a field trip may also benefit from SH-TRACE-based networks. An interesting application that fits very well to an SH-TRACE-based network is communication among a group of hearing disabled people who communicate with sign language. Since vision is the only possible means of communication for such a group, without direct vision (*i.e.*, you cannot see simultaneously a person at your left and another at your right), it is not possible to have group communication in all situations. If each person has a Personal Digital Assistant (PDA) with a small camera and a low-resolution monitor large enough to display the signs, possibly with several panels, and a Motion Picture Expert Group (MPEG) coder [127], which enables high compression, then it is possible to create a communication network for hearing disabled people.

The remainder of this chapter is organized as follows. Section 6.2 describes the SH-TRACE protocol in detail. Section 6.3 provides analysis of the performance of SH-TRACE and simulations to compare SH-TRACE with other MAC protocols. Section 6.4 gives some discussion of the features of SH-TRACE, and Section 6.5 presents a summary of the chapter.

## 6.2 Protocol Architecture

### 6.2.1 Overview

SH-TRACE is an energy-efficient dynamic TDMA protocol designed for real-time data broadcasting. In SH-TRACE, data transmission takes place according to a dynamically updated transmission schedule. Initial access to data slots are through contention, but once a node reserves a data slot, its reservation for a data slot in the subsequent frames continues automatically as long as the node continues to broadcast a packet in each frame. Thus nodes only need to contend for data slots at the beginning of data bursts.

A controller in the network is responsible for creating the TDMA schedule based on which nodes have continued reservations from previous frames and which nodes have successfully contended for data slots in the current frame. The controller transmits this schedule to the rest of the nodes in the network at the beginning of the data sub-frame. Whenever the energy of the controller drops below the energy level of the other nodes in the network by more than a set amount, it assigns another radio with higher energy than itself as the next controller. Controller handover takes place during the TDMA schedule transmission by specifying the ID of the new controller.

Finally, if the number of transmissions in a frame exceeds a predetermined threshold, each node listens only to data from certain nodes. Each node determines which transmitters to listen to based on information obtained from all the nodes during the information summarization (IS) slots. The following sub-sections describe these ideas in more detail.

### 6.2.2 Basic Operation

SH-TRACE is organized around time frames with duration matched to the periodic rate of voice packets. The frame format is presented in Figure 6.2. Each frame consists of two sub-frames: a control sub-frame and a data sub-frame. The control sub-frame consists of a beacon slot, a contention slot, a header slot, and an IS slot.

At the beginning of every frame, the controller node transmits a beacon message. This is used to synchronize all the nodes and to signal the start of a new frame. The contention slot, which immediately follows the beacon message, consists of $N_C$ sub-slots. Upon hearing the beacon, nodes that have data to send but did not reserve data slots in the previous frame, randomly choose sub-slots to transmit their requests. If the contention is successful (*i.e.*, no collisions), the controller grants a data slot to the contending node. The controller then sends the header, which includes the data transmission schedule of the current frame. The transmission schedule is a list of nodes that have been granted data slots in the current frame along with their data slot numbers. A contending node that does not hear its ID in the schedule understands that its contention was unsuccessful (*i.e.*, a collision occurred or all the data slots are already in use) and contends again in the following frame. If the waiting time for a voice packet during contention for channel access exceeds the threshold, $T_{drop}$, it is dropped. The header also includes the ID of the controller for the next frame, which is determined by the current controller according to the node energy levels.

The IS slot begins just after the header slot and consists of $N_D$ sub-slots. Nodes that are scheduled to transmit in the data sub-frame transmit a short IS message exactly in the same order as specified by the data transmission schedule. An IS message includes the energy level of the transmitting node, enabling the controller node to monitor the energy level of the entire network, and an end-of-stream bit, which is set to one if the node has no data to send. Each receiving node records the received power level of the transmitting node and inserts this information into its IS table. The information in the IS table is used as a proximity metric for the nodes (*i.e.*, the higher the received power the shorter the distance between transmitter and receiver nodes). Using the receive signal strength to estimate the relative distance of the transmitter to the receiver is a method employed in previous studies [38, 15]. If the number of transmissions in a particular frame is higher than a predetermined number of transmissions, $N_{MAX}$, each node schedules itself to wake up for the top $N_{MAX}$ transmissions that are the closest transmitters to the node. Hence the network is softly partitioned into many virtual clusters based on the receivers; this is

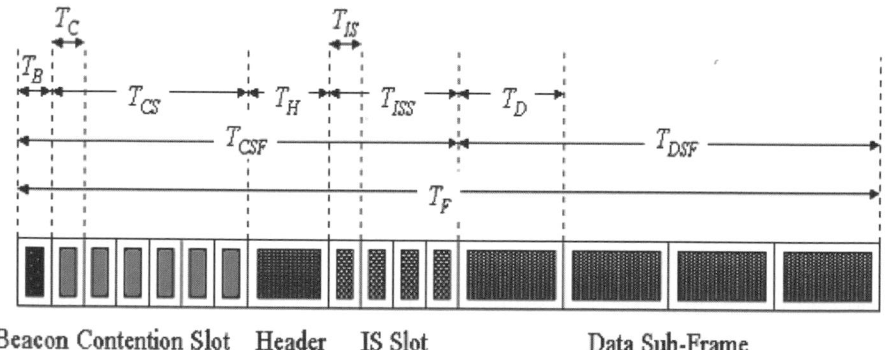


*Figure 6.2.* SH-TRACE frame format.

fundamentally different from transmitter based network partitioning. Note that other methods of deciding which nodes to listen to can be used within the SH-TRACE framework by changing what data nodes send in the IS slot.

The data sub-frame is broken into constant length data slots. Nodes listed in the schedule in the header transmit their data packets at their reserved data slots. Each node listens to at most $N_{MAX}$ data transmissions in a single frame; therefore each node is on for at most $N_{MAX}$ data slots. All nodes are in the sleep mode after the last reserved data slot until the beginning of the next frame.

If the power level of the controller node is lower than any other node by a predetermined threshold, then in the next frame controller handover takes place. The controller node assigns another node (any other node in the network with energy level higher than that of the controller) as the controller, effective with the reception of the header packet. Upon receiving the header packet, the node assigned to be the controller assumes the controller duties.

A node keeps a data slot once it is scheduled for transmission as long as it has data to send. A node that sets its end-of-stream bit to one because it has no more data to send will not be granted channel access in the next frame (*i.e.*, it should contend to get a data slot once it has new data to send). Automatic renewal of data slot reservation enables real-time data streams to be uninterrupted [6].

### 6.2.3 Initial Startup

At the initial startup stage, a node listens to the medium to detect any ongoing transmissions for one frame time $T_F$, because it is possible that there might already be an operational network. If no transmission is detected, then the node picks a random time, smaller than the contention slot duration $T_{CS}$, at which to transmit its own beacon signal, and the node listens to the channel until its contention timer expires. If a beacon is heard in this period, then the node stops its timer and starts normal operation. Otherwise, when the timer expires, the node sends a beacon and assumes the controller position. In case there is a beacon collision, none of the colliding nodes will know it, but the other nodes hear the collision, so the initial startup continues. All the previously collided nodes, and the nodes that could not detect the collision(s) because of capture, will learn of the collisions with the first successful beacon transmission.

### 6.2.4 Prioritization

SH-TRACE supports an optional prioritized operation mode. In this mode, the nodes have three pre-assigned priority levels, of which Priority Level-1 ($PL1$) is the highest priority and $PL3$ is the lowest priority. The highest level has the highest quality of service (QoS), and the lowest level has the lowest QoS. Prioritization is incorporated into the basic protocol operation at three points: contention, scheduling, and receiver based soft clustering. In the contention stage, $PL1$, $PL2$, and $PL3$ nodes have $N_{C1}$, $N_{C2}$, and $N_{C3}$ number of non-overlapping contention slots, respectively. $N_{Ci}$ is chosen to satisfy

$$\frac{N_{PL1}}{N_{C1}} < \frac{N_{PL2}}{N_{C2}} < \frac{N_{PL3}}{N_{C3}} \tag{6.1}$$

where $N_{PLi}$ denotes the expected number of nodes in priority level $i$. The number of contention slots per node is higher for the higher priority levels, which results in less contention for higher priority nodes.

In scheduling, $PL1$ and $PL2$ nodes are always given channel access, even if all the data slots are reserved. If all the data slots are reserved, then reservations of $PL3$ nodes are canceled starting from the latest reservation and granted to the higher priority nodes.

All the nodes should listen to data from PL1 nodes, whether or not they are close to the nodes. Prioritization does not affect the general protocol operation, because we assume that the number of $PL1$ and $PL2$ nodes is much less than the number of $PL3$ nodes.

### 6.2.5 Receiver-Based Soft Cluster Creation

Each node creates its own receiver-based listening cluster, which has a maximum of $N_{MAX}$ members, by choosing the closest nodes based on the proximity information obtained from the received power from the transmissions in the IS slot. Priority has precedence over proximity; therefore, transmissions by $PL1$ nodes are always included in the listening cluster by removing the furthest node in the cluster. To avoid instantaneous changes in the listening clusters and to make them more stable, there is also a continuity rule: a member of the listening cluster cannot be excluded from the listening cluster until it finishes its talk spurt, which is a natural extension in the sense that if a speech stream is broken in the middle, the whole transmission becomes useless.

### 6.2.6 Reliability

In case the controller node fails, the rest of the network should be able to compensate for this situation and should be able to continue normal operation as fast as possible. Failure of the controller manifests itself at two possible points within a frame: Beacon transmission and header transmission. A backup controller, assigned by the controller, could listen for the beacon and header and become the controller whenever the controller fails. However, if both the backup controller and the controller die simultaneously, then the network is left dead. Instead of assigning a backup controller, there is a more natural and complete way of backing up the network: the transmission schedule is a perfect list of backup controllers in a hierarchical manner. The first node in the schedule is the first backup controller, the second node is the second backup controller, and the $N$'th node is the $N$'th backup controller.

The backup nodes listen to the beacon, which is a part of normal network operation. If the first backup controller does not hear the beacon for *Inter Frame Space* ($T_{IFS}$) time, then the controller is assumed dead and the first node transmits the beacon. If the beacon is not transmitted for a certain time ($2T_{IFS}$), then the second backup controller understands that both the controller and the first backup controller are dead, and transmits the beacon. The backup procedure works in the same way for all the nodes listed in the transmission schedule in the previous frame. If after $(N + 1)T_{IFS}$ time no beacon is transmitted, then the rest of the nodes understand that the controller and all the backup nodes are dead, and they restart the network. Re-startup is the same as the initial network startup, but in this case nodes do not listen for an existing controller for $T_F$; instead they start right away, because they know the controller is dead and there is no need for waiting.

The response of the network to controller failure in header transmission is very similar to that of beacon failure. The succeeding backup node transmits the transmission schedule of the previous frame by updating it with the information in the IS slot of the previous frame denoting nodes with reservations that no longer have data to transmit. However, none of the nodes, including the backup nodes, listen to the contention slot, so the transmission schedule cannot be updated for the contending nodes. This is not much of an issue in voice transmission, because packet loss due to delayed channel access causes the early packets to be dropped, which is preferable over packet loss in the middle of a conversation [6]. Since controller node failure is not a frequent event, it is better not to dissipate extra energy on controller backup. If all the backup nodes die simultaneously during header transmission, then the rest of the nodes begin re-startup. Also if there were no transmissions in the previous frame, then in case of a controller failure, nodes just enter re-startup (*i.e.*, there are no backup nodes).

## 6.3 Simulations and Analysis

To test the performance of SH-TRACE, we conducted simulations using the ns-2 software package [128]. We simulated conversational voice coded at 32 Kbps. The channel rate is chosen as 1 Mbps. We used a perfect channel without any loss or error models. Each node listens to a maximum of 5 nodes (*i.e.*, $N_{MAX} = 5$). The transport agent used in the simulations is UDP, which is a best effort service. All the simulations, unless otherwise stated, are run for 100 s and averaged for 3 independent runs. We used the Hierarchical Reference Point Group Mobility (HRPGM) as our mobility model, which is introduced in Section 2.4. The maximum distance between the nodes is 250 m in the scenarios we employed, and maximally separated nodes could hear each

*Table 6.1.* Parameters used in the SH-TRACE simulations.

| Acronym | Description | Value |
|---|---|---|
| $T_F$ | Frame Duration | 25 ms |
| $T_{CSF}$ | Contention sub-frame duration | 3.8 ms |
| $T_{DSF}$ | Data sub-frame duration | 21.2 ms |
| $T_B$ | Beacon duration | 40 $\mu s$ |
| $T_{CS}$ | Contention slot duration | 2.32 ms |
| $T_C$ | Contention sub-slot duration | 40 $\mu s$ |
| $T_H$ | Header slot duration | 440 $\mu s$ |
| $T_{ISS}$ | IS slot duration | 1 ms |
| $T_{IS}$ | IS sub-slot duration | 40 $\mu s$ |
| $T_D$ | Data slot duration | 848 $\mu s$ |
| $T_{IFS}$ | Inter-frame space duration | 16 $\mu s$ |
| $T_{drop}$ | Packet drop threshold | 50 ms |
| $N_D$ | Number of data slots | 25 |
| $N_C$ | Number of contention sub-slots | 58 |
| $N_{Ci}$ | Number of contention sub-slots in priority $i$ | 3, 5, 50 |
| $N_{MAX}$ | Maximum listening cluster size | 5 |
| $P_T$ | Transmit power | 600 mW |
| $P_R$ | Receive power | 300 mW |
| $P_I$ | Idle power | 100 mW |
| $P_S$ | Sleep power | 0 mW |
| $m_s$ | Average spurt duration | 1 s |
| $m_g$ | Average gap duration | 1.35 s |
| $D_{Tr}$ | Transmit range | 250 m |
| $D_{CS}$ | carrier Sense range | 507 m |

other's transmissions. Acronyms, descriptions and values of the parameters used in the simulations are presented in Table 6.1.

### 6.3.1 Frame Structure and Packet Sizes

Frame time, $T_F$, is chosen to be 25 ms, which is the periodic rate of voice packet generation; of this 25 ms, 21.2 ms is for the data sub-frame, $DSF$, and 3.8 ms is for the control sub-frame, $CSF$. There are 58 40 $\mu s$ duration contention sub-slots, 25 40 $\mu s$ duration IS sub-slots, and 25 848 $\mu s$ duration data slots. The number of contention slots is approximately equal to $e$ times the number of data slots, because the optimal throughput of a Slotted ALOHA system is $1/e$. Beacon, contention, and IS packets are all 3 bytes. The header packet has a variable length of 3-53 bytes, consisting of 3 bytes of packet header and 2 bytes of data for each node to be scheduled. The data packet is 104 bytes long, consisting of 4 bytes of packet header and 100 bytes of data. Variations in the packet sizes are due to the differences in the information content of each packet. Each slot or sub-slot includes 16 $\mu s$ of guard band ($IFS$) to account for switching time and round-trip time.

### 6.3.2 Voice Source Model

In voice source modeling, we assume each node has a voice activity detector, which classifies speech into "spurts" and "gaps" (*i.e.*, gaps are the silent moments during a conversation) [5–7]. During gaps no data packets are generated, and during spurts data packets are generated in the rate of the speech coder, which is 32 Kbps in our simulations. Both spurts and gaps are exponentially distributed statistically independent random variables, with means $m_s$ and $m_g$, respectively. In our simulations and analysis we used the experimentally verified values of $m_s$ and $m_g$, which are 1.0 s and 1.35 s, respectively [7].

### 6.3.3 Throughput

A maximum of 25 nodes can transmit data simultaneously; therefore, the maximum achievable total throughput is 800 Kbps. However, it is not possible to reach this upper bound while ensuring that QoS is met. QoS in the context of voice traffic corresponds to the packet drop ratio, $R_{PD}$, due to the packet delay exceeding a certain maximum delay, $T_{drop}$ ($T_{drop}$ = 50 ms). $R_{PD}$ is the ratio of the average number of dropped voice packets per frame and the average number of voice packets generated per frame. Since the voice signals are composed of spurts and gaps, it is possible to support more than 25 users by multiplexing more than 25 conversational speech sources into 25 data slots.

Figure 6.3 shows a plot of the average number of data packets generated per frame as a function of the number of nodes in the network. The theoretical value of the average number of data packets generated per frame, $N_G$, in a network of $N_N$ nodes is obtained as

$$N_G = \frac{m_s}{m_s + m_g} N_N \tag{6.2}$$

Both theoretical and simulation curves increase linearly with almost constant slope with $N_N$. All the simulation data points are within 3.0% error range of the theoretical curve, with a maximum difference of 0.85 packets per frame at $N_N = 60$. Figure 6.3 shows that the average number of voice packets generated per frame is 43% of the number of voice sources.

It is possible to achieve a normalized capacity, $\eta$, of 2.35 conversations per channel with perfect multiplexing of the voice sources over time, which means that SH-TRACE can theoretically support a maximum of 58 nodes with no packet drop. The normalized capacity is defined in [7] as the ratio of the maximum number of nodes (*i.e.*, conversations) that can be supported without exceeding the packet drop ratio of 0.01 and the number of channels (data slots). However, the voice sources are independent (*i.e.*, they are not coordinated, as the input pattern is not a design parameter), and it would be too optimistic to expect perfect statistical multiplexing. Therefore, we expect packet drops to occur with fewer than 58 nodes.

The theoretical average number of packets delivered per frame, $N_A$, is obtained as:

$$N_A = min\left[\frac{m_s}{m_s + m_g} N_N, N_{DS}\right] \tag{6.3}$$

where $N_{DS}$ is the total number of data slots in a frame (25 in our simulations). Curves showing the average number of delivered packets per frame obtained from the simulations and theory are in good agreement for $N_N < 50$ (see Figure 6.3). However, for $N_N \geq 50$ the difference between the curves is large (*i.e.*, at $N_N = 60$ the difference is 2.1 packets per frame). In theory we did not consider any packet drops, and we assumed data packets are distributed evenly in all frames. In simulations, both of these assumptions are violated for $N_N > 50$. For $N_N > 58$, the average number of packets per frame exceeds the number of data slots; because of this, in our theoretical model $N_A = 25$, but we cannot achieve this upper bound in the simulations. This is because of the fact that in some frames the number of voice packets are smaller than 25, and

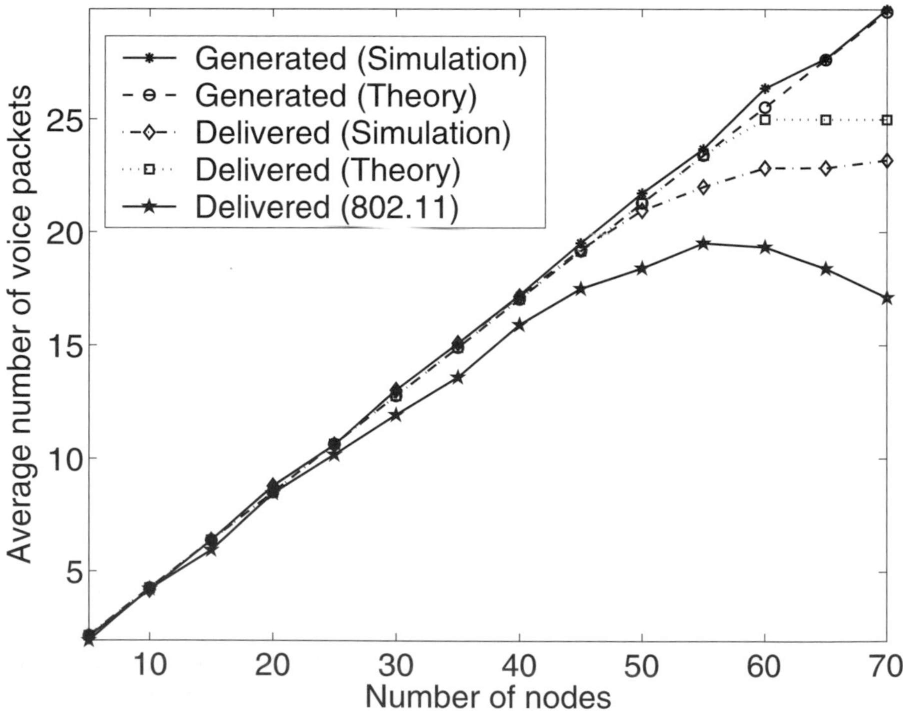


*Figure 6.3.* Average number of voice packets per frame vs. total number of nodes with active voice sources.

in some others much higher than 25. Thus, due to the independent statistical behavior of the voice sources, it is not possible to achieve the upper bound without sacrificing QoS (*i.e.*, $R_{PD}$).

Figure 6.3 also shows the number of data packets delivered per $T_F$ time for IEEE 802.11, which is lower than that of SH-TRACE for all $N_N$. The maximum difference between SH-TRACE and IEEE 802.11 is 6.1 packets per $T_F$ time at $N_N = 70$, which corresponds to a 26.2% decrease in throughput.

For broadcast traffic, IEEE 802.11 does not use the standard four-way handshake mechanism; instead only the data packet is transmitted, since no feedback can be obtained from the other nodes, and Binary Exponential Backoff (BEB) is not employed for broadcast traffic [20]. Thus IEEE 802.11 becomes Carrier Sense Multiple Access (CSMA) for broadcast traffic [19]. The throughput of IEEE 802.11 is lower than SH-TRACE due to collisions, which arise because of the lack of coordination among the nodes (*i.e.*, simultaneous transmissions result in collisions and none of the transmitting nodes are aware of the situation).

Figure 6.4 shows the average number of packets delivered per frame per node as a function of the number of nodes in the network. For $N_N < 40$, the nominal value, 0.43, is preserved, but for larger numbers of nodes, per node capacity starts to decrease exponentially. The nominal value of average number of data packets delivered per frame per node is given as: $\frac{m_s}{m_s+m_g}$, which is 0.43. With the increasing number of data packets and in the absence of perfect multiplexing, the voice packets are not distributed evenly among the frames. Thus packets exceeding $T_{drop}$ are automatically dropped, which is the main contributor to the per node capacity decrease. However, for $N_N > 58$, even if there were perfect multiplexing, packet drops are unavoidable because after that point the average number of data packets per frame exceeds the number of data slots.

Figure 6.5 illustrates a particular example of TRACE operation for a network with 50 nodes. Figure 6.5 (a) shows the number of voice packets generated per frame as a function of time. Although the average number of voice packets per frame is 21.26, the number of voice packets generated during a given frame exceeds the maximum capacity, 25, frequently, which results in packet drops. Figure 6.5 (b) and Figure 6.5 (c) display the number of dropped packets per frame and the number of collisions per frame for the voice traffic shown in Figure 6.5 (a), respectively. The average number of dropped packets per frame and the average number of collisions per frame are 0.63 and 0.024, respectively. Thus, while theoretically the network should be able to handle the traffic from 50 nodes with no data loss, the offered traffic sometimes exceeds the network capacity (25 data slots) and packets must be dropped.

Figure 6.6 shows the average number of dropped packets per frame and $R_{PD}$ as functions of $N_N$ in the upper and lower panels, respectively. $R_{PD}$ increases exponentially for $N_N \geq 40$. In this range, the actual number of nodes that simultaneously have voice packets to send frequently exceeds the number of data slots, so voice packets are dropped since it is not possible to grant permission to all nodes simultaneously.

The normalized capacity, $\eta$, of SH-TRACE reaches 1.76 at $N_N = 44$ ($R_{PD} = 0.01$), whereas the $\eta$ of PRMA is reported as 1.16 [7]. It is also reported in [7] that at an optimal operating point the of PRMA reaches 1.64. However, the problem of keeping the network in the optimal operating point is not addressed in [7]. So the $\eta$ at the optimal case can be thought of as the upper bound for PRMA. There are several factors contributing to the difference between the $\eta$'s of PRMA and SH-TRACE. The main factor in this difference is that the contention for channel access results in collisions and data slots cannot be used by either of the contenders in PRMA. In SH-TRACE, since contention is not in the data slots, there is no loss of data slots due to contention. In addition, the number of contention slots is higher than the number of data slots, which further reduces the collisions. Another factor is that the $T_{drop}$ of PRMA is 20% lower than that of SH-TRACE.

Channel bit rate used in [6, 7] for PRMA evaluation is 720 Kbps, which is entirely used by the nodes for uplink communications. The bandwidth used by the controller for downlink communications is not mentioned in [6, 7]. We

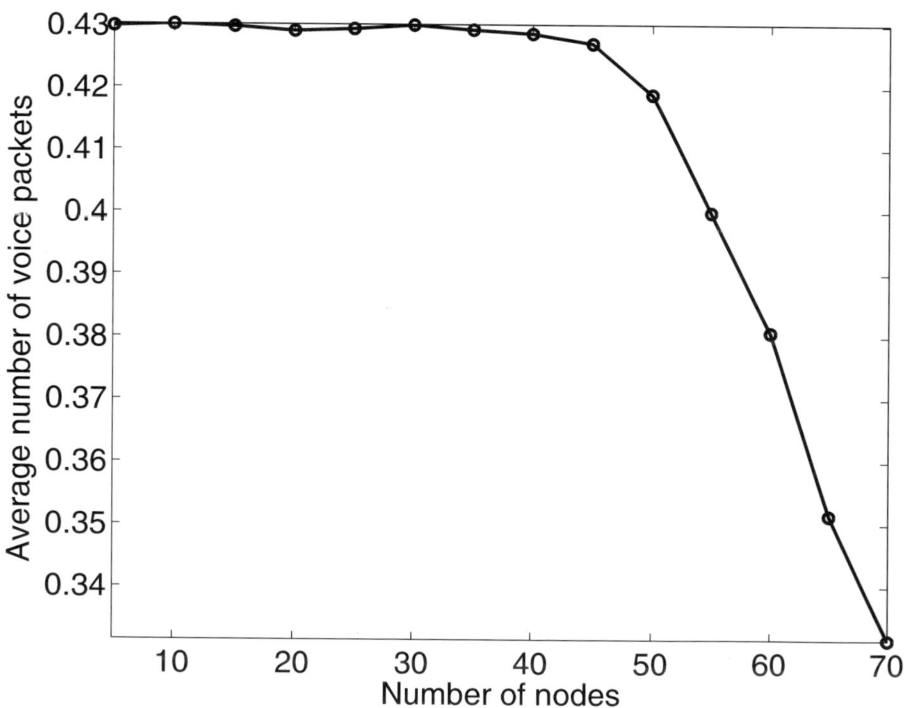


*Figure 6.4.* Average number of voice packets delivered per frame per node vs. number of nodes.

used a channel bit rate of 1 Mbps, which includes both uplink and downlink bandwidth and all the control packets. The bandwidth exclusively used for data transmissions and receptions is 848 Kbps.

### 6.3.4 Energy Dissipation

The energy dissipation in the network is due to transmit, receive and idle modes of the radio and can be written as

$$E = E_T + E_R + E_I \tag{6.4}$$

where $E$, $E_T$, $E_R$, and $E_I$ are total energy dissipation, energy dissipated for transmission, energy dissipated for reception, and idle energy dissipation, respectively. All the energy values are the averages for a single frame duration. Acronyms and descriptions of the variables are given in Table 6.2.

Total transmit energy dissipation is given by

$$E_T = E_B^T + E_C^T + E_H^T + E_{IS}^T + E_D^T \tag{6.5}$$

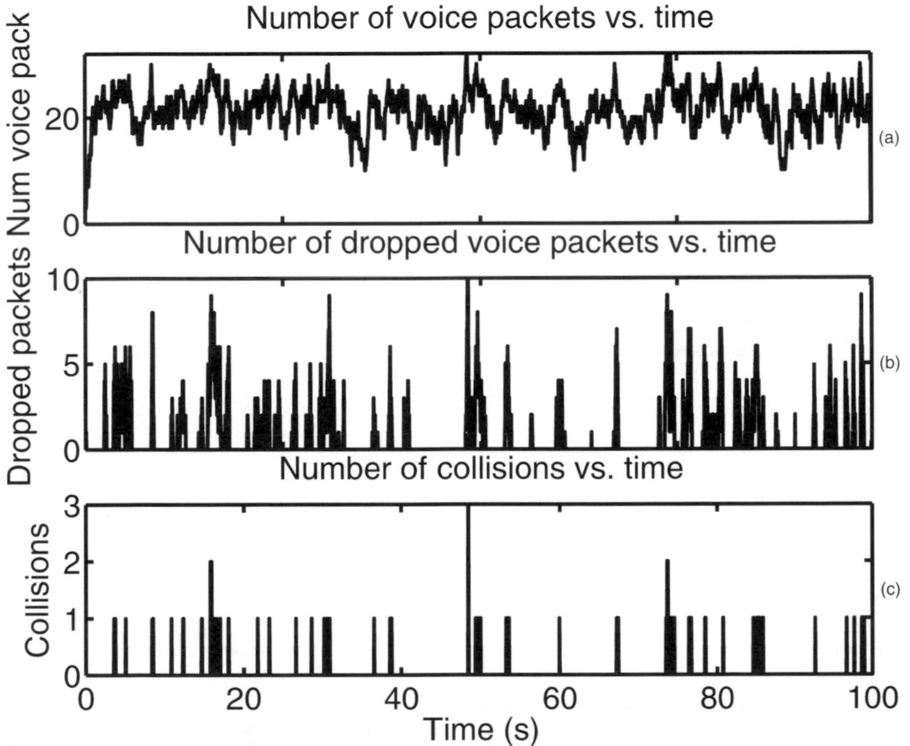


*Figure 6.5.* (a) Actual number of voice packets generated per frame as a function of time with $N_N = 50$ and $N_A = 21.26$. (b) Number of dropped packets per frame for the voice traffic in (a). (c) Number of collisions per frame for the same traffic.

where $E_B^T$, $E_C^T$, $E_H^T$, $E_{IS}^T$, and $E_D^T$ are beacon contention, header, IS, and data transmission energy dissipations, respectively. Energy dissipated for beacon transmission in terms of beacon duration, $T_B$, and transmit power, $P_T$, is given by

$$E_B^T = T_B P_T \tag{6.6}$$

Energy dissipation for contention is similar to beacon transmission, but the average number of contentions per frame is a statistical quantity. We define the following parameters: the average data burst duration, $T_{DB}$, which is the average length of a data burst (*i.e.*, average duration of a speech burst, $m_s$), the average silence time between data bursts, $T_S$, (*i.e.*, average gap duration, $m_g$), the contention packet duration, $T_C$, the average number of data packets per frame, $N_A$, and, the frame duration, $T_F$. Using this notation, the contention energy dissipation per frame is given as

$$E_C^T = N_A \frac{T_F}{T_{DB} + T_S} T_C P_T \tag{6.7}$$

In the above equation we assumed all data bursts need to contend once to gain access to the channel (*i.e.*, there are no collisions). This is a reasonable

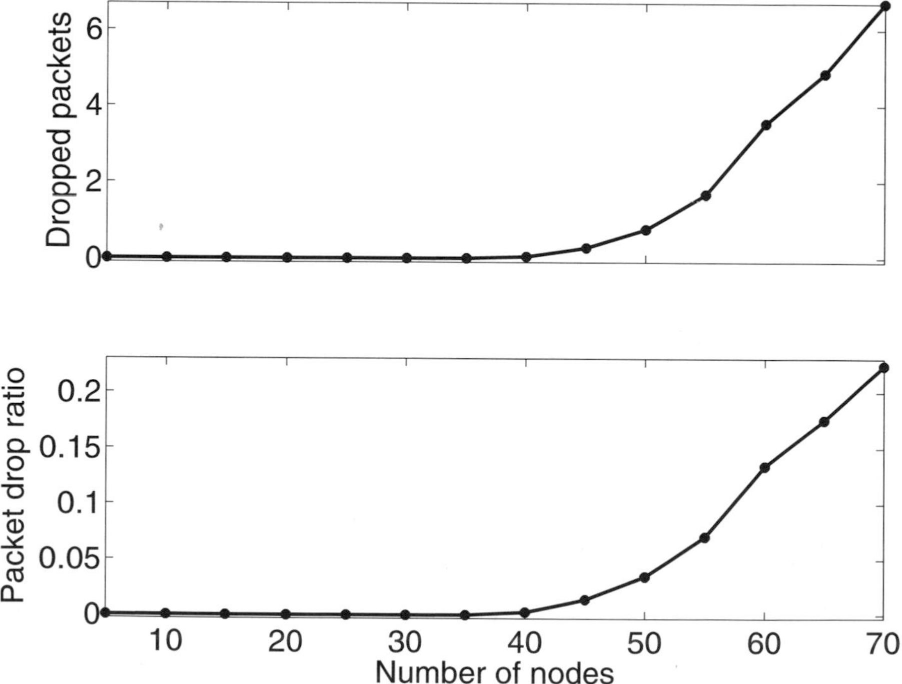


*Figure 6.6.* The upper panel displays the average number of dropped packets per frame as a function of $N_N$, and the lower panel displays the average value of packet drop ratio, $R_{PD}$.

*Table 6.2.* Acronyms and descriptions of the variables used in the energy calculations.

| Acronym | Description |
|---|---|
| $E$ | Total energy dissipation per frame |
| $E_T$ | Transmit energy dissipation per frame |
| $E_R$ | Receive energy dissipation per frame |
| $E_I$ | Idle mode energy dissipation per frame |
| $E_B^T$ | Energy dissipation for beacon transmission per frame |
| $E_C^T$ | Energy dissipation for contention packet transmission per frame |
| $E_H^T$ | Energy dissipation for header transmission per frame |
| $E_{IS}^T$ | Energy dissipation for IS transmission per frame |
| $E_D^T$ | Energy dissipation for data transmission per frame |
| $E_B^R$ | Energy dissipation for beacon reception per frame |
| $E_C^R$ | Energy dissipation for contention reception per frame |
| $E_H^R$ | Energy dissipation for header reception per frame |
| $E_{IS}^R$ | Energy dissipation for IS reception per frame |
| $E_D^R$ | Energy dissipation for data reception per frame |

assumption, because the number of contention slots is large enough to generally avoid collisions, and while there are still a small number of collisions, this does not affect our analysis significantly.

The header is a variable length packet consisting of constant overhead and a variable payload that is a function of $N_A$

$$E_H^T = T_H(N_A)P_T \tag{6.8}$$

$T_H(N_A)$ is the duration of the header as a function of $N_A$

$$T_H(N_A) = T_{OH} + N_A T_{DP} \tag{6.9}$$

where $T_{OH}$ is the time spent for overhead and $T_{DP}$ is the time spent to schedule one data packet.

Energy spent for IS transmission can be expressed in terms of $N_A$, $P_T$, and IS packet duration, $T_{IS}$,

$$E_{IS}^T = N_A T_{IS} P_T \tag{6.10}$$

Energy dissipation for data transmission is similar to IS transmission

$$E_D^T = N_A T_D P_T \tag{6.11}$$

where $T_D$ is the duration of the data packet.

Energy dissipated for data reception can be decomposed into beacon reception, $E_B^R$, contention reception, $E_C^R$, header reception, $E_H^R$, IS reception, $E_{IS}^R$, and data reception, $E_D^R$, components; hence, the total receive energy dissipation is

$$E_R = E_B^R + E_C^R + E_H^R + E_{IS}^R + E_D^R \tag{6.12}$$

All the nodes, except the controller, receive the beacon at the beginning of each frame, independent of data traffic. Energy dissipated for beacon reception can be written in terms of the number of nodes in the network, $N_N$, the time for the beacon, $T_B$ and the receive power, $P_R$,

$$E_B^R = (N_N - 1)T_B P_R \tag{6.13}$$

Contention packets are received by the controller only. Thus the expression for contention reception energy dissipation is the same as the contention transmission, except in this case we use $P_R$ instead of $P_T$

$$E_C^R = N_A \frac{T_F}{T_{DB} + T_S} T_C P_R \tag{6.14}$$

Energy dissipation for header reception is

$$E_H^R = (N_N - 1)T_H(N_A)P_R \tag{6.15}$$

IS packets have constant duration, $T_{IS}$, and they are received by all nodes, and transmitted by all nodes that are scheduled to transmit data. Thus the energy to receive IS packets is:

$$E_{IS}^R = (N_N - 1)N_A T_{IS} P_R \tag{6.16}$$

All the nodes in the network listen to a maximum of $N_{MAX}$ transmissions; in a situation where $N_A$ is smaller than $N_{MAX}$, then only $N_A$ transmissions are received. Therefore, data reception energy dissipation is

$$E_D^R = N_N min[N_{MAX}, N_A] T_D P_R \tag{6.17}$$

Idle energy dissipation is mainly dominated by the controller. The controller is on for the whole contention slot, which is transmission free for most of the time. The idle energy expression in terms of idle power, $P_I$, total contention slot length, $T_{CS}$, and the other previously defined parameters is

$$E_I = (T_{CS} - N_A \frac{T_F}{T_{DB} + T_S} T_C)P_I \tag{6.18}$$

Figure 6.7 shows a plot of the total network energy dissipation per frame for different values of $N_N$. Theoretical analysis and simulation results are in good agreement, with a maximum difference of 4.0 mJ (3.7%) when $N_N = 60$. The difference arises due to the overestimation of $N_A$. In theory, we did not consider the packet dropping probability; however, starting with $N_N = 40$, there is a non-zero packet dropping probability. Nonetheless, the energy mismatch between the theory and simulation is still small (3.7% maximum). The theoretical minimum energy is the energy needed to transmit and receive data only. We assume an omniscient network controller takes care of network coordination and informs the nodes without dissipating any energy. The maximum difference between the theoretical minimum and the simulation results is 19.6 mJ (15.8%) at $N_N = 70$. All the energy above the theoretical minimum energy is spent for control packets and network monitoring.

Energy dissipation without the IS slot is much higher than energy dissipation when the IS slots are used to create listening clusters, because all the nodes should be listening to all data transmissions, forwarding the desired packets to the upper layer and discarding the rest, which results in extra power dissipation for unnecessary but also inevitable information reception in the absence of the IS slot. The maximum difference between the case without the IS slot and with the IS slot is 335 mJ, which corresponds to a 269% increase in energy dissipation. Thus using data summarization slots (IS slots) are very helpful in reducing energy dissipation.

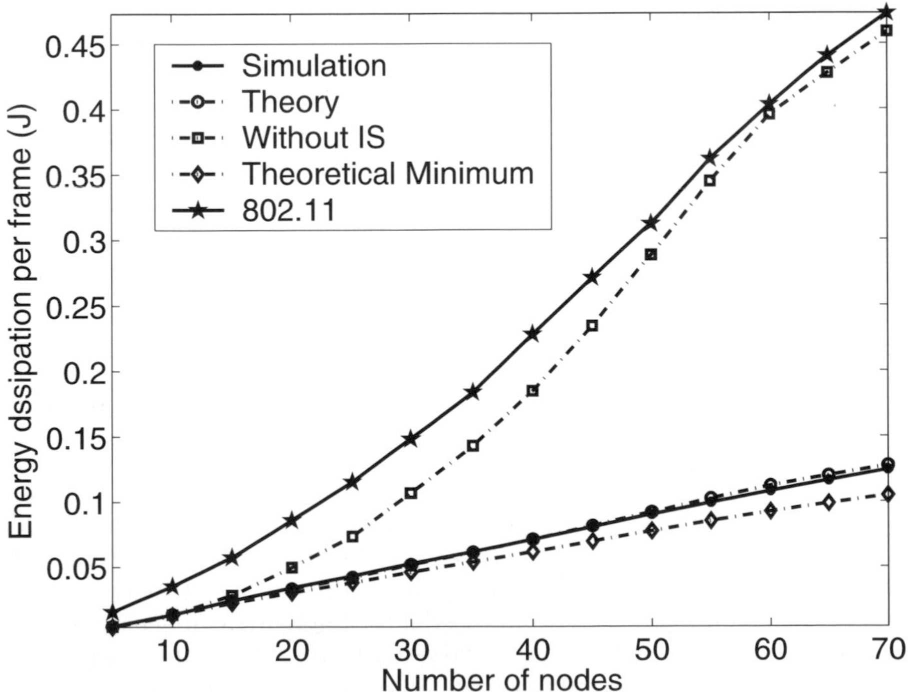


*Figure 6.7.* Average network energy dissipation per frame vs. number of nodes.

IEEE 802.11 has 52 bytes of packet header in broadcast packets in standard operation, whereas SH-TRACE has only 4 bytes of data packet header. In order to compare these two protocols on a fair basis, we reduced the header size for IEEE 802.11 to 4 bytes, so the data packet size is 104 bytes for both SH-TRACE and IEEE 802.11 in our simulations. Figure 6.7 shows that energy dissipation for IEEE 802.11 is higher than all the other cases for all $N_N$, because in standard IEEE 802.11 operation all the nodes in the network are always on and all the broadcast packets are received without any discrimination. The maximum difference between SH-TRACE and IEEE 802.11 energy dissipation curves is 349 mJ (281% increase in energy dissipation) at $N_N = 70$. Energy dissipation for IEEE 802.11 is higher than that of SH-TRACE without IS slots because in IEEE 802.11, none of the nodes goes to sleep mode, whereas in SH-TRACE without IS slots, nodes go to sleep mode if the network is idle.

Figure 6.8 (a), (b), and (c) show the energy dissipation per node per frame in transmit, receive, and idle modes for SH-TRACE and IEEE 802.11, respectively. IEEE 802.11 has almost constant transmit energy dissipation at all node densities, because all the packets are transmitted in IEEE 802.11 without being dropped. Transmit energy of SH-TRACE is almost constant and higher than that of IEEE 802.11 for $N_N < 60$, due to additional control packet transmissions. However, for $N_N \geq 60$, due to the dropped packets, transmit energy dissipation of SH-TRACE is lower than that of IEEE 802.11. Receive energy dissipation of SH-TRACE is constant for $N_N \geq 15$, after which the average number of transmissions exceeds the maximum listening cluster size. IEEE 802.11 receive energy increases linearly with node density until $N_N = 60$, and stays constant for $N_N \geq 60$. Idle energy dissipation of SH-TRACE is almost zero for all node densities. IEEE 802.11 idle energy dissipation decreases with increasing node density, because idle time is decreasing with increasing node density, as transmit and receive time are increased.

Total energy dissipation per node per frame for SH-TRACE and IEEE 802.11 at $N_N = 5$ are 0.83 mJ and 3.19 mJ, respectively. The ratios of transmit, receive, and idle energy dissipation at $N_N = 5$ for SH-TRACE and IEEE 802.11 are 1.0 / 2.46 / 0.22 and 1.0 / 2.39 / 11.17, respectively. Energy dissipation of SH-TRACE and IEEE 802.11 for packet transmission and reception are almost the same, because the listening cluster ($N_{MAX} = 5$) does not save any energy at this node density for SH-TRACE. Most of the extra energy dissipation for IEEE 802.11 when compared to SH-TRACE is due to the idle mode energy dissipation, which constitutes 73% of the total energy dissipation. At $N_N = 70$, the per node per frame energy dissipation for SH-TRACE and IEEE 802.11 are 1.83 mJ. and 6.96 mJ, respectively. The ratios of transmit, receive, and idle energy dissipation at $N_N = 70$ for SH-TRACE and IEEE 802.11 are 1.0 / 8.71 / 0.03 and 1.0 / 27.51 / 2.55, respectively. The difference between SH-TRACE and IEEE 802.11 is mostly due to the listening cluster based power

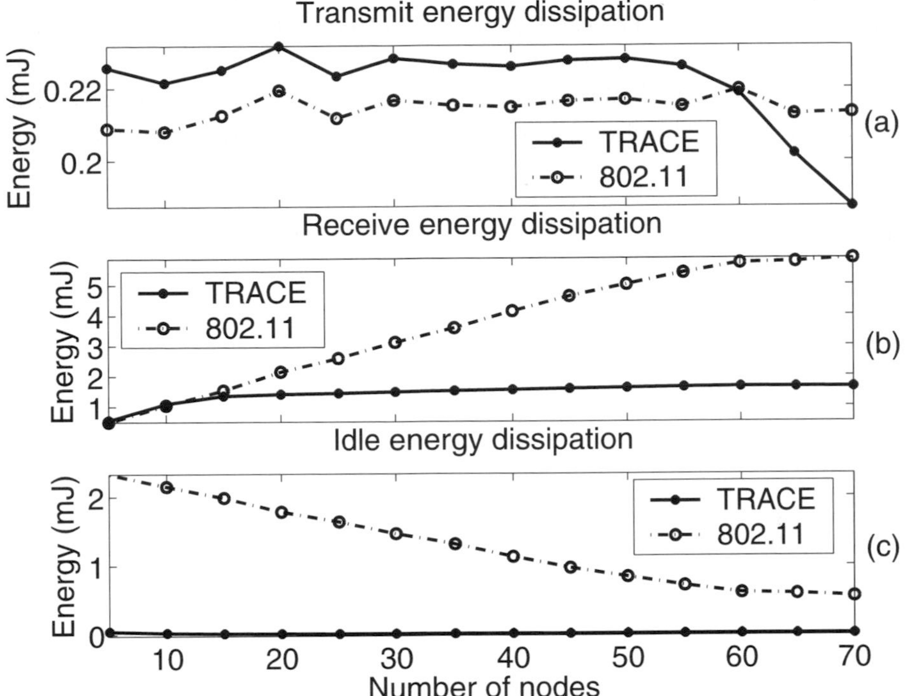


*Figure 6.8.* (a) Transmit energy dissipation per node per frame for SH-TRACE and IEEE 802.11. (b) Receive energy dissipation per node per frame for SH-TRACE and IEEE 802.11. (c) Idle energy dissipation per node per frame for SH-TRACE and IEEE 802.11.

saving mechanism of SH-TRACE, because most of the energy dissipation of IEEE 802.11 (*i.e.*, 85% of total energy dissipation) is due to the packet receptions at this node density.

Energy dissipation is a function of data traffic, which is directly proportional to the number of nodes. For lower node densities, the dominant factor in energy dissipation for IEEE 802.11 is idle listening. Thus, if the idle power and sleep power are very close in an energy model, then the energy dissipation for SH-TRACE and IEEE 802.11 will be very close in a low density network. If the node density is high, then the dominant term in energy dissipation for IEEE 802.11 is the receive power, and the contribution of idle mode energy dissipation becomes marginal.

### 6.3.5 Packet Delay

The arrival time of a voice packet is uniformly distributed to one frame time. It is not possible for a packet to arrive and be delivered in the same frame; the earliest delivery can be in the next frame. The delivery time is a uniform discrete random variable, because packets can be delivered only at the end of each data slot, and no data slot has precedence over others.

Random variables $x$ and $y$, which are shown in Figure 6.9, represent the packet arrival time and the packet delivery time, respectively. The probability density function (pdf) of $x$, the packet arrival time, is given as

$$f_x(x) = \begin{cases} 1/T_F & 0 < x \leq T_F \\ 0 & otherwise \end{cases} \tag{6.19}$$

The pdf of the delivery time, $y$, is

$$f_y(y) = \frac{1}{N_A} \sum_{k=1}^{N_A} \delta(y - T_{CSF} + kT_D) \tag{6.20}$$

where $T_{CSF}$ is the control sub-frame duration, and $\delta(.)$ is the Dirac-delta function.

We can find the delay by subtracting $x$ from $y$, but we must add an offset of $T_F$ to $y$ in order to define both variables according to beginning of frame 1 (*i.e.*,

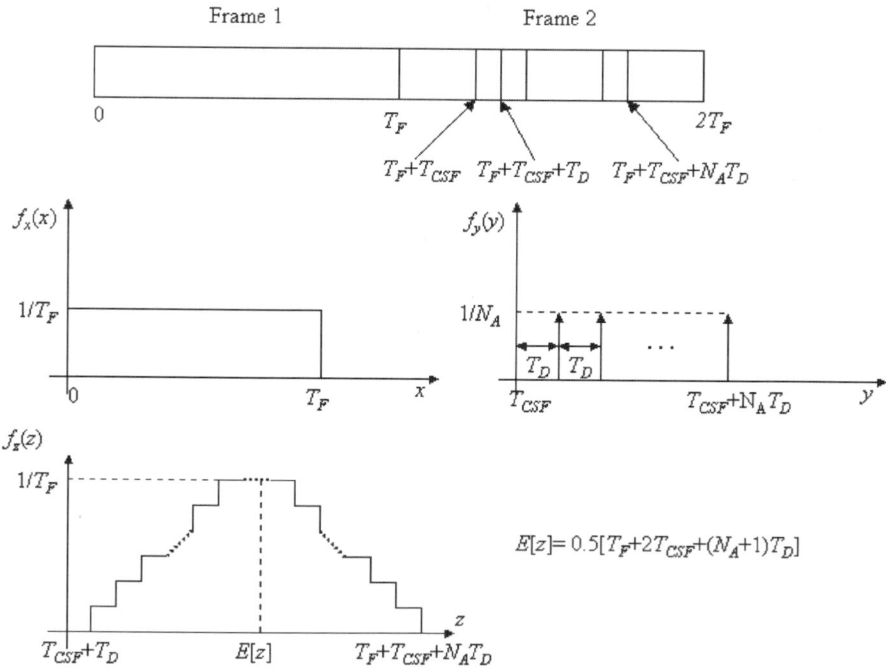


*Figure 6.9.* Packet delay calculations. The top row displays the frame structure used for packet delay analysis. The pdf's of x, y, and z are plotted in middle and bottom rows.

$y = 0$ corresponds to $y = T_F$). The delay is given by

$$z = T_F + y - x \tag{6.21}$$

Since $x$ is a uniform random variable between 0 and $T_F$, $T_F - x$ is equivalent to $x$, so

$$z = y + x \tag{6.22}$$

The pdf of $z$ is obtained by convolving the probability density functions (pdfs) of $x$ and $y$

$$f_z(z) = f_x(x) \bigotimes f_y(y) \tag{6.23}$$

$$f_z(z) = \frac{1}{N_A T_F} \sum_{k=1}^{N_A} \left\{ \begin{array}{c} u(z - (T_{CSF} + kT_D)) \\ -u(z - (T_F + T_{CSF} + (N_A + 1 - k)T_D)) \end{array} \right\} \tag{6.24}$$

where $u(.)$ denotes the unit step function. The expected value of $z$ is obtained as

$$E[z] = 0.5(T_F + 2T_{CSF} + (N_A + 1)T_D) \tag{6.25}$$

Figure 6.10 shows a plot of the pdfs obtained from simulation and theory. Root mean square (RMS) error between the two curves is less than 0.2%. Figure 6.11 shows a plot of the average packet delay versus the number of nodes. The maximum difference between the simulation data and theory is 0.26 ms at $N_N = 70$, which corresponds to a 1.0% difference.

### 6.3.6 Node Failure

To test the automatic controller backup scheme, we designed a random controller failure simulation. In the simulation the controller can fail with a probability p at each frame. This corresponds to an exponentially decreasing nonfailure probability in time, which is shown to be a valid model for wireless radios [129]. Let $u$ be the random variable that represents the non-failure for the controller at the $k$'th beacon transmission and define $q = 1 - p$ to be the probability of non-failure. The pdf of $u$ is

$$f_u(k) = \left(\frac{1-q}{q}\right) q^k \tag{6.26}$$

The first term is the normalization term to make the area of the pdf unity; the second term states that the probability of non-failure decreases exponentially. The expected value of $u$ is

$$\mu = \left(\frac{1-q}{q}\right) \sum_{k=0}^{\infty} q^k \tag{6.27}$$

This gives the average lifetime (*i.e.*, failure time) of a network without any backup mechanism and with a controller non-failure probability of $q$. The expected lifetime of a network having a backup mechanism with $N$ nodes, $\mu_N$, is given by

$$\mu_N = N\mu \tag{6.28}$$

Network lifetime curves obtained from simulations and theory with $p = 0.1$ are plotted in Figure 6.12. Simulations are averaged over 10 statistically independent simulation runs. The average network lifetime without backup is 0.2824 s and 0.2778 s for the simulation and theory, respectively. The average network lifetime with backup elongates the network failure time directly proportional with the number of nodes in the network. Network lifetime increases 50 times for a 50-node network theoretically. The increase in network lifetime in the simulations is 52.4, on the average for a 50-node network.

One of the design goals in the controller failure monitoring and compensation is to enable the network to resume its normal operation in an uninterrupted manner. We found that the data packet per frame per node is an appropriate metric to test the continuity of the normal network operation (*i.e.*, since the

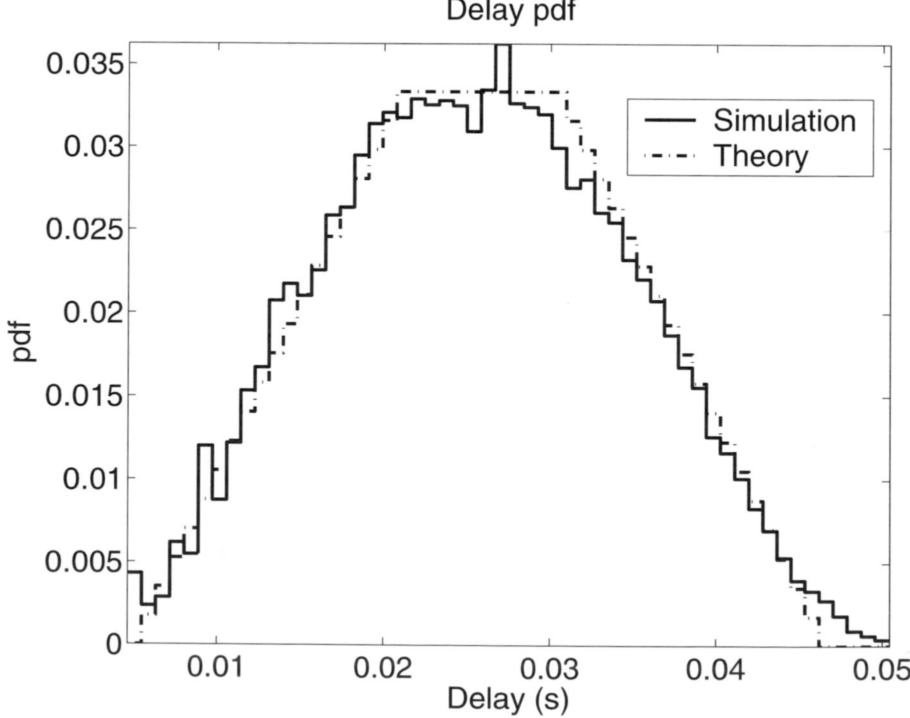


*Figure 6.10.* Pdf of packet delay with $N_N = 50$. RMS error between the simulation and theory is 0.16%.

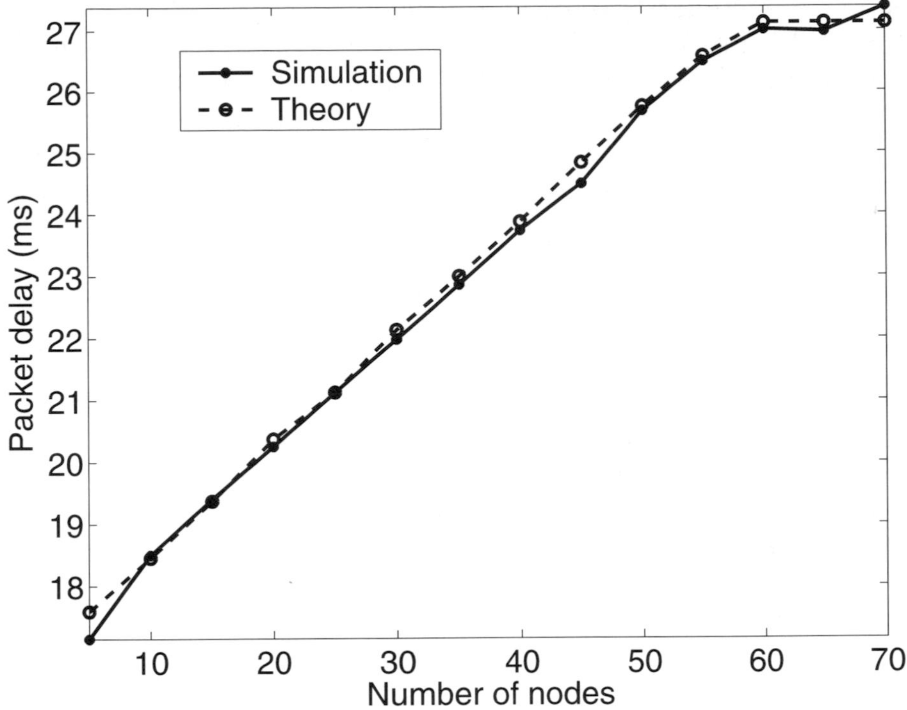


*Figure 6.11.* Packet delay vs. number of nodes.

nodes keep dying, the total number of nodes and consequently the number of transmitted data packets are reduced proportional to this decrease in the number of alive nodes). We also set $m_g = 0$, so that each alive node in the network has a data packet at each frame and the statistical behavior of the voice source does not interfere with our metric (*i.e.*, as an alive node might not have data to send in the actual voice model in all frames, then it would not be possible to quantify the behavior of the network correctly). In Figure 6.13 we present curves showing the average number of received data packets per frame per node as a function of time for a 20-node network assuming no node failures (dashed line) and for the same network with node failures and the backup mechanism turned on (solid line). Data per frame per node is equal to unity for both curves for the whole simulation time during which there is at least one alive node left for the case with node failures (*i.e.*, $t < 6.4s$), which shows that the backup mechanism can effectively compensate for the controller failure, and until all the nodes die the network continues to operate with minimal interruption in service.

### 6.3.7 Virtual Cluster Smoothing

Figure 6.14 shows a plot of the number of node changes in the virtual clusters per node per frame with and without the continuity rule for a 50-node network. The differences between the curves arise due to the fact that without the continuity rule a continuing voice stream is dropped because a closer voice source starts to transmit its voice packets. The total number of changes in the virtual clusters without the continuity rule is 48,639, whereas it is 42,813 with the continuity rule, which shows a 12% reduction in the total number of changes. In other words, 5826 voice burst interruptions are prevented from happening by applying the continuity rule.

### 6.3.8 Priority Levels, Dropped Packets, and Collisions

In the simulations, almost all the dropped data packets are from $PL3$ nodes. There were very few dropped packets at $PL1$ or $PL2$ nodes, and very few collisions of contention packets from these nodes. As long as the number of voice packets is below the number of data slots for a particular frame, the number of collisions and the number of packet drops are virtually zero. The

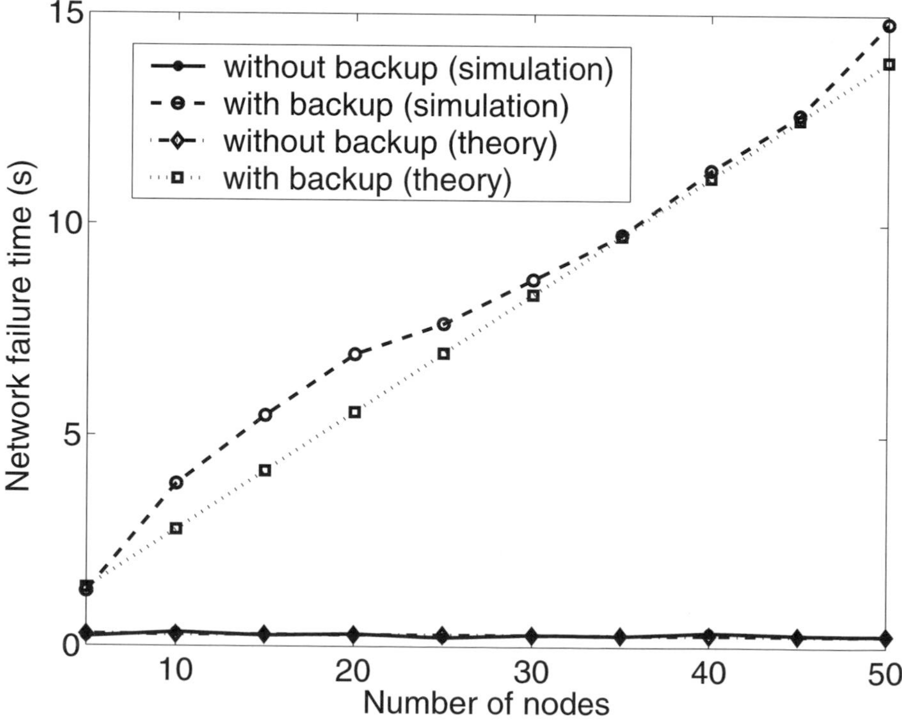


*Figure 6.12.* Network failure time vs. number of nodes.

$R_{PD}$ is non-zero for $N_N \geq 40$ (see Figure 6.6) because of the fact that nodes attempting to get channel access are unable to get access for several frames due to temporary overload. Nodes that cannot obtain channel access continue contention until they get channel access, which results in an increased number of contending nodes, and more collisions. This also explains why there are very few packet drops for $PL1$ and $PL2$ nodes: since there is no congestion for high priority nodes, they get channel access in a single attempt, and the number of contending nodes does not increase even in overloaded traffic. Statistical multiplexing of voice packets is good enough to ensure high QoS for high priority nodes (*i.e.*, if all the high priority nodes try to get channel access at the same frame, there would be a non-negligible collision probability. Since we observed only a few collisions, we conclude that statistical multiplexing is good enough to avoid collisions for high priority nodes.). For low priority nodes, there is not much contention except for overloaded traffic frames, which also reinforces our observation about the statistical multiplexing of voice packets.

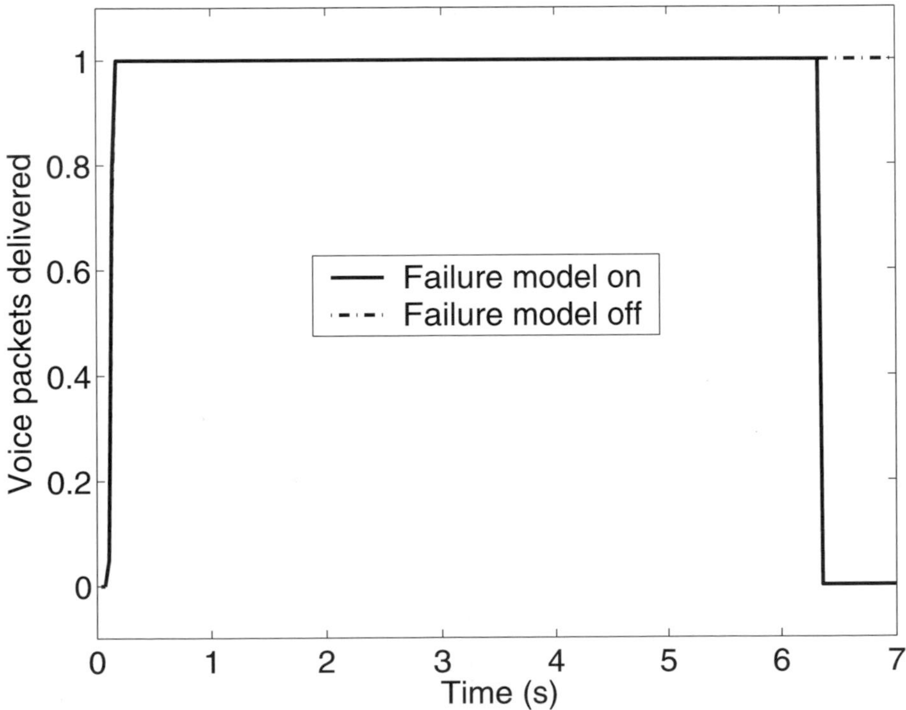


*Figure 6.13.* Delivered voice packets per frame per alive node vs. time.

## 6.4 Discussion

In the simulations we assumed that all the nodes in the network are active voice sources and independent of each other to demonstrate the worst-case performance of SH-TRACE; however, it is unlikely in a realistic scenario that everybody is speaking without listening to others. Therefore, it is possible to support a higher number of nodes with the same packet drop rate in a realistic scenario. Energy dissipation per node will also be lower if not all the nodes are active. There will not be any change in packet delay characteristics, because silent nodes are just passive participants in the network.

We consider the possibility of saving more energy by using a multi-hop approach, but it turns out that since the dominant term in our radio model is the energy dissipation on radio electronics, we cannot save any power by a multi-hop approach with the radio model and coverage area we are using.

Capture is a factor that affects the fairness of PRMA and all other ALOHA family protocols. Indeed, a strong capture mechanism increases the throughput of PRMA, because of the fact that most of the contention attempts result in favor of the nodes close to the base station. Instead of loosing both packets

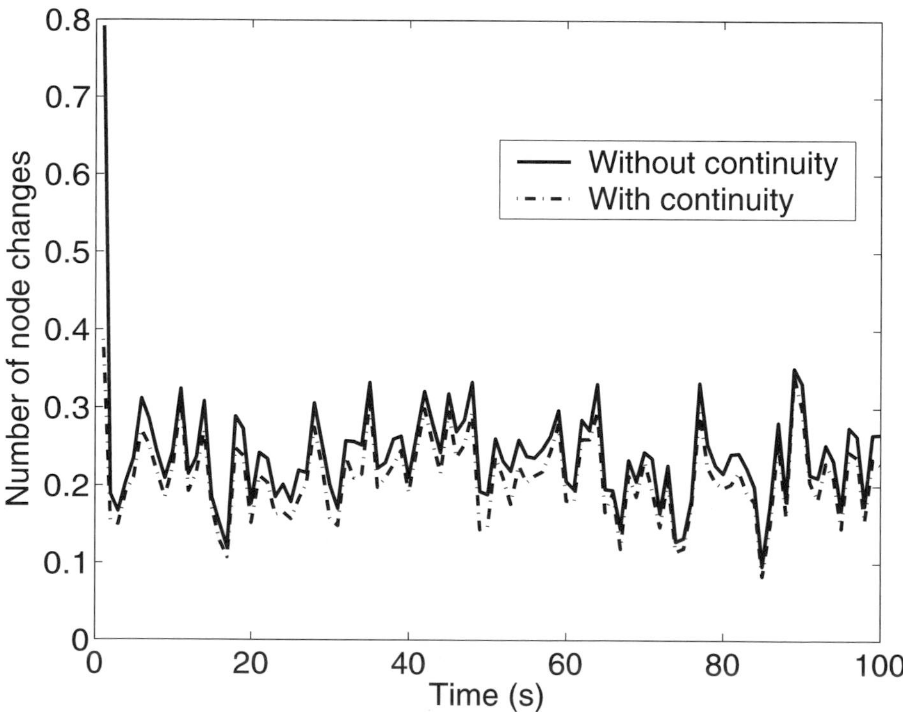


*Figure 6.14.* Average number of node changes in listening clusters per node per frame as a function of time.

and wasting the whole data slot, only one of the nodes looses the contention and the other captures the channel, which increases the total throughput and degrades the fairness among the nodes in an uncontrolled manner (*i.e.*, unlike the prioritization in SH-TRACE, which is a controllable design parameter). The effects of capture in SH-TRACE are only marginal.

The IS slot contributes significantly to the energy efficiency of SH-TRACE. The end-of-stream information is included in the IS slot, because it is the most appropriate point in the frame structure for this information. A node does not know whether it has a voice packet or not in the next frame during its data transmission because the packet generation rate is matched to the frame rate, so end-of-stream information cannot be sent in the data slot. The earliest point where a node knows it is out of packets is during the control sub-frame. If the end-of-stream information is not sent in the IS slot but in the data slot (*i.e.*, no data is sent to indicate the end-of-stream like in PRMA), then the controller should be listening to all the data slots to monitor for the continued use of data slots, which results in waste of considerable energy.

In our current implementation, the information for data discrimination is proximity; however, the information in the IS slot can be modified for different applications. For example, the IS slot can be used to send metadata describing the data that will be transmitted in the corresponding data slot. The nodes can choose which transmitters to listen to based on this metadata. An efficient way of using metadata prior to data transmission in a multi-hop sensor network application is presented in [52].

Priority levels of SH-TRACE might be used to support various requirements of the applications using SH-TRACE as the MAC layer. For example, in a military operation, it is necessary that the commander has priority over other soldiers and everybody listens to the commander's speech ($PL1$), and the leaders of each sub-squad should also have a priority lower than that of the commander ($PL1$) but higher than the others ($PL3$). In a multimedia application $PL1$ and $PL2$ could be thought of as Constant Bit Rate (CBR) sources and $PL3$ as Variable Bit Rate (VBR) or Available Bit Rate (ABR) sources. In a field trip, the tour guide can be a $PL1$ node and the rest of the group members can be $PL3$ nodes.

SH-TRACE does not have a global synchronization requirement. Each node updates the frame start time by listening to the beacon sent by the controller, and all the transmissions and receptions are defined with respect to this time, which is updated at each frame by the controller.

SH-TRACE is virtually immune to stability problems because the contention is not in the data slots but in contention sub-slots. The natural isolation between the contention-free data sub-frame and the contention sub-slots makes SH-TRACE highly stable and robust.

A comparison of an early version of SH-TRACE, PBP (an enhanced version of IEEE 802.11 for single-hop networks) and ASP (an energy-efficient polling protocol for Bluetooth) in a sensor network application for a many-to-one data transmission model is given in [120]. It is shown that the energy dissipation of SH-TRACE is much less than PBP for the same number of data transmissions. PBP is shown to be not very energy-efficient when compared to SH-TRACE because of the lack of central coordination and high overhead.

The channel utilization mechanism of SH-TRACE is designed for data sources that produce data packets periodically with data bursts spanning several frame times. However, channel utilization for data traffic will suffer seriously due to the non-bursty nature of data packets. As a remedy to this problem, we present a multi-stage contention algorithm in Appendix A.

## 6.5 Summary

In this chapter, we describe SH-TRACE in detail and evaluate its performance through computer simulations and theoretical analysis. SH-TRACE is a time frame based MAC protocol designed primarily for energy-efficient reliable real-time voice packet broadcasting in a peer-to-peer, single-hop infrastructureless radio network. Such networks have many application areas for various scenarios that obey a strongly connected group mobility model, such as interactive group trips, small military or security units, and mobile groups of hearing impaired people. SH-TRACE is a centralized MAC protocol that separates contention and data transmission, providing high throughput, low delay and stability under a wide range of data traffic. Furthermore, SH-TRACE uses dynamic scheduling of data transmissions and data summarization prior to data transmission to achieve energy efficiency, which is crucial for battery-operated lightweight radios. In addition, energy dissipation is evenly distributed among the nodes by switching network controllers when the energy from the current controller is lower than other nodes in the network, and reliability is achieved through automatic controller backup features. SH-TRACE can support multiple levels of QoS, and minimum bandwidth and maximum delay for voice packets are guaranteed to be within certain bounds.

Chapter 7

# MH-TRACE PROTOCOL ARCHITECTURE

## 7.1 Introduction

In Chapter 6 we presented SH-TRACE, which is an energy-efficient QoS supporting reliable MAC protocol for fully connected ad hoc networks. However, due to limited radio range, barriers, and interference it is not possible to restrict a communication network to a fully-connected topology. Although for the application scenarios considered in Chapter 6 users need to communicate with their immediate (*i.e.*, single-hop) neighbors, a multi-hop extension of the SH-TRACE protocol to support single-hop communications within a multi-hop (*i.e.*, not fully connected) network topology is necessary. Furthermore, this is the logical next step to pave the way for energy-efficient, QoS supporting, multi-hop, real-time data broadcast, multicast, and unicast routing.

In this chapter, we present the Multi-Hop Time Reservation using Adaptive Control for Energy efficiency (MH-TRACE) protocol architecture for energy-efficient single-hop voice broadcasting in a multi-hop network [42, 43, 45]. Ad hoc network architectures for mobile radios have many application areas in several scenarios that involve groups of people. Examples of such groups are military units (*e.g.*, a squadron of soldiers), search and rescue teams, and tourists in interactive group trips. The ad hoc network architecture for these applications should be capable of supporting broadcasting of real-time traffic like voice, which is the primary means of conveying information in interactive human groups. To support such real-time broadcast traffic, the network protocol must provide support for quality of service (QoS), such as bounding delay and reducing packet drops. Furthermore, the network protocol should avoid unnecessary energy dissipation, since light-weight mobile radios are battery operated and have limited energy.

This chapter is organized as follows. Section 7.2 describes the MH-TRACE protocol in detail. Section 7.3 provides analysis of the performance of MH-TRACE and simulations to compare MH-TRACE with other MAC protocols. Section 7.4 gives some discussion of the features of MH-TRACE, and Section 7.5 summarizes the chapter.

## 7.2 Protocol Architecture

### 7.2.1 MH-TRACE Operation

Figure 7.1 shows a snapshot of MH-TRACE clustering and medium access for a portion of an actual distribution of mobile nodes. In MH-TRACE, the network is organized into overlapping clusters through a distributed algorithm. Section 7.2.3 explains the details of the cluster creation and maintenance algorithms. Time is organized around superframes with duration, $T_{SF}$, matched to the periodic rate of voice packets, where each superframe consists of $N_F$ frames. The frame format is presented in Figure 7.2. Each frame consists of two sub-frames: a control sub-frame and a data sub-frame. The control sub-frame consists of a beacon slot, a clusterhead announcement (CA) slot, a contention slot, a header slot, and an information summarization (IS) slot. Acronyms and descriptions of MH-TRACE specific terms are presented in Table 7.1.

At the beginning of each occupied frame, the clusterhead transmits a beacon message. This is used to announce the existence and continuation of the cluster to the cluster members and the other nodes in the transmit range of the clusterhead. By listening to the beacon and CA packets, all the nodes in the carrier sense range of this clusterhead update their interference level table. Each clusterhead chooses the least noisy frame to operate within and dynamically changes its frame according to the interference level of the dynamic network. Collisions with the members of other clusters are minimized by the clusterhead's selection of the minimal interference frame.

The contention slot, which immediately follows the CA slot, consists of $N_C$ sub-slots. Upon hearing the beacon, each node that has data to send but did not reserve a data slot in the previous cyclic superframe, randomly chooses a subslot to transmit its request. If the contention is successful (*i.e.*, no collisions), the clusterhead grants a data slot to the contending node.

Following the contention subslot, the clusterhead sends the header, which includes the data transmission schedule of the current frame. The transmission schedule is a list of nodes that have been granted data slots in the current frame, along with their data slot numbers. A contending node that does not hear its ID in the schedule understands that its contention was not successful (*i.e.*, a collision occurred or all the data slots are already in use) and contends again in the following superframe. If the waiting time for a voice packet during contention for channel access exceeds the threshold, $T_{drop}$, the packet is dropped.

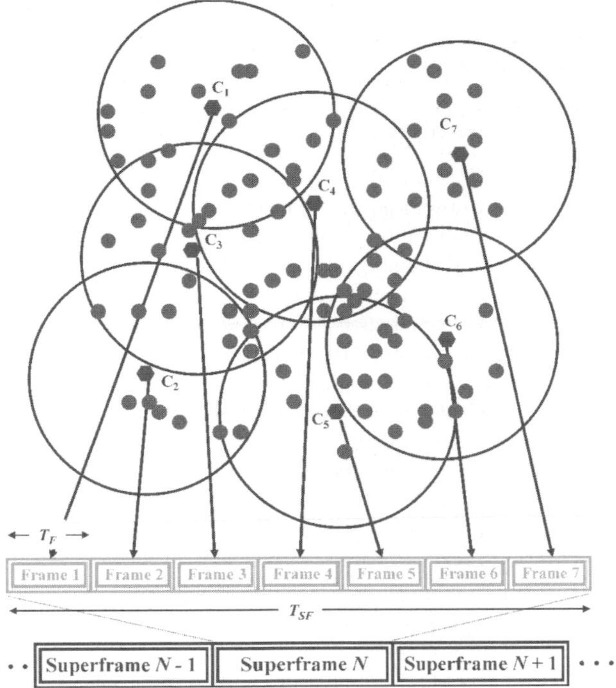


*Figure 7.1.* A snapshot of MH-TRACE clustering and medium access for a portion of an actual distribution of mobile nodes. Nodes $C_1$ through $C_7$ are clusterhead nodes.

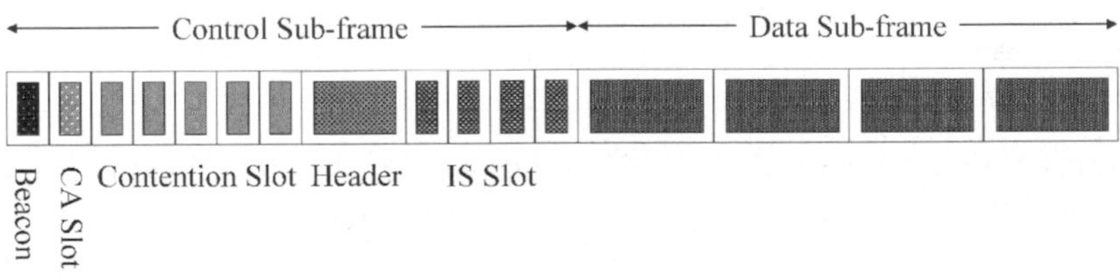


*Figure 7.2.* MH-TRACE frame format.

The information summarization (IS) slot begins just after the header slot and consists of $N_D$ sub-slots. Each node that is scheduled to transmit data sends a short IS packet prior to actual data transmission exactly in the same order as specified by the data transmission schedule. Based on these IS packets, neighbor nodes decide whether to stay awake and receive the data packets or enter the sleep mode for the duration of the data packet and avoid reception of irrelevant or collided data packets. An IS packet includes the ID of the transmitting node and an end-of-stream bit, which is set to one if the node has no data to send.

*Table 7.1.* MH-TRACE acronyms, descriptions, and values.

| Acronym | Description | Value |
|---|---|---|
| CH | Clusterhead | NA |
| CA | Clusterhead Announcement | NA |
| IF | Information Summarization | NA |
| $N_N$ | Total number of nodes in the network | 50-200 |
| $T_{SF}$ | Superframe duration | 25.172 ms |
| $T_F$ | Frame duration | 3.596 ms |
| $T_B$ | Beacon slot duration | 32 $\mu s$ |
| $T_{CA}$ | CA slot duration | 32 $\mu s$ |
| $T_C$ | Contention sub-slot duration | 32 $\mu s$ |
| $T_H$ | Header slot duration | 92 $\mu s$ |
| $T_{IS}$ | IS sub-slot duration | 92 $\mu s$ |
| $T_D$ | Data slot duration | 432 $\mu s$ |
| $T_{IFS}$ | Inter-frame space duration | 16 $\mu s$ |
| $T_{drop}$ | Packet drop threshold | 50 ms |
| $N_F$ | Number of frames within superframe | 7 and 10 |
| $N_{MAX}$ | Maximum listening cluster size | 5 |
| $P_T$ | Transmit power | 600 mW |
| $P_R$ | Receive power | 300 mW |
| $P_I$ | Idle power | 100 mW |
| $P_S$ | Sleep power | 0 mW |
| $D_{Tr}$ | Transmit range | 250 m |
| $D_{CS}$ | carrier Sense range | 507 m |
| $p_{CA}$ | CA transmission probability | 0.5 |
| $p_{CF}$ | Frame change probability | 0.5 |

Each receiving node records the received power level of the transmitting node and inserts this information into its IS table. The IS table is used as a proximity metric for the nodes. Nodes that are not members of this cluster also listen to the IS slot and record the received power level. Each node creates its own listening cluster by selecting the top $N_{MAX}$ transmissions that are the closest transmitters to the node. Note that other methods of deciding which nodes to listen to can be used within the MH-TRACE framework by changing what data nodes send in the IS slot (in our implementation there is no information about the data, such as metadata summarizing the data content, or transmitting node, such as priority). Hence the network is softly partitioned into many virtual clusters (called listening clusters) based on the receivers. Section 8.2.6 further elaborates on listening cluster creation.

The data subframe is broken into constant length data slots. Nodes listed in the schedule in the header transmit their data packets at their reserved data slots. A node keeps a data slot once it is scheduled for transmission as long as it has data to send, which enables real-time data streams to be uninterrupted. A node that sets its end-of-stream bit (in the IS packet) to one because it has no more data to send will not be granted channel access in the next superframe.

### 7.2.2 Energy Savings Techniques

There are two techniques used in MH-TRACE to save energy. The first technique is to reduce energy dissipation at the MAC layer. Nodes should be in sleep mode whenever possible to avoid (*i*) dissipating energy in the idle state, (*ii*) overhearing transmissions initiated from nodes that are further than the successful transmission range (*i.e.*, carrier sensing), and (*iii*) receiving corrupted packets due to collisions.

Any node in the startup mode cannot enter the sleep mode until it reaches the steady-state mode. If a node either transmitted (clusterhead node) or received (non-clusterhead node) a header packet within $2T_{SF}$ time, it is in steady-state mode. Otherwise, it is in startup mode. Similarly, all nodes are required to be awake for all Beacon, CA and IS slots for all the frames within the superframe to gather the control information to run MH-TRACE seamlessly. Ordinary nodes also stay awake to receive the header slot of their own clusterhead. In addition, clusterheads stay awake in their own frames through the contention slot to receive any contention requests. Figure 7.3 illustrates the sleep/active states of MH-TRACE.

The second technique is to reduce energy dissipation by avoiding packet receptions that will be discarded at the higher layers of the protocol stack if not avoided at the MAC layer. Based on the information sent in the IS slots, the MAC layer can decide whether or not to receive the data packets. If there is no discrimination of packets and all packets are to be received, then each node stays awake for all the data transmissions in its receive range, and goes to sleep mode in the data slots that are known to be empty or result in collisions through listening to the IS slots. Thus, traffic adaptive energy efficiency is achieved even without data discrimination. However, by employing data discrimination through listening cluster creation, further energy savings can be achieved. In the simulations we used proximity, which is obtained from the receive power of the IS packets, as our discrimination metric and set a maximum size, $N_{MAX}$, on the number of listening cluster members.

### 7.2.3 MH-TRACE Clustering

Unlike existing clustering approaches [88, 92, 130, 98], the MH-TRACE clustering scheme is not based on connectivity information, which can be gathered by sacrificing some of the bandwidth to disseminate and collect the $k$-hop connectivity information. Almost all of the existing clustering algorithms create a unique clustering for a given node distribution; thus they are deterministic.

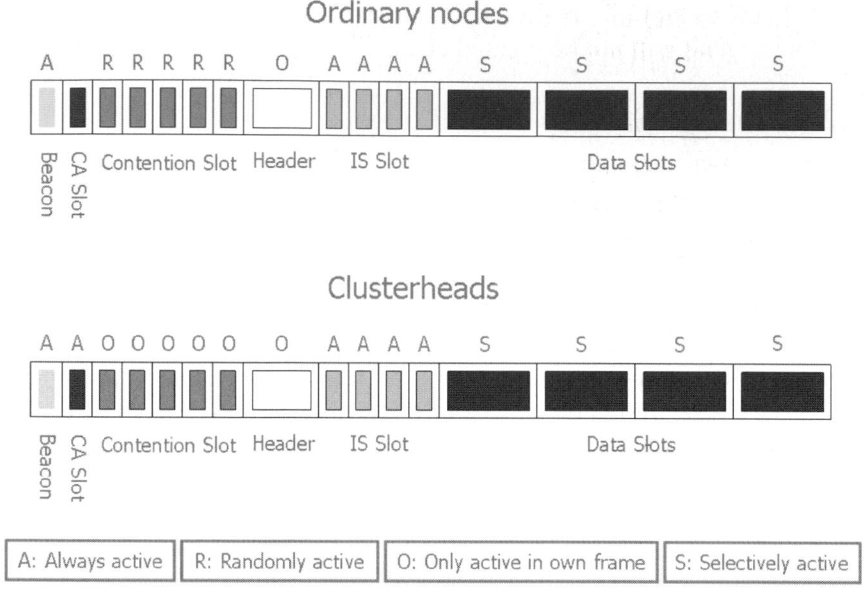


*Figure 7.3.* MH-TRACE sleep/active states.

In MH-TRACE cluster creation and maintenance, the overhead is lower when compared to the other clustering approaches, because the only information a node needs to know in order to form a cluster is the interference level in the different time-frames, which is monitored continuously to minimize the interference between clusters. However, for a given node distribution there are many clustering possibilities in MH-TRACE; thus it is probabilistic. By using the interference level as a constraint for cluster creation, secondary effects, like inter-cluster interference, are also incorporated into cluster creation, which is crucial in avoiding collisions. Interference is not considered as a constraint in the other clustering approaches.

Instead of frequency division or code division, MH-TRACE clusters use the same spreading code or frequency, and inter-cluster interference is avoided by using time division among the clusters to enable each node in the network to receive all the desired data packets in its receive range, not just those from nodes in the same cluster. Thus, our clustering approach does not create hard clusters—the clusters themselves are only used for assigning time slots for nodes to transmit their data.

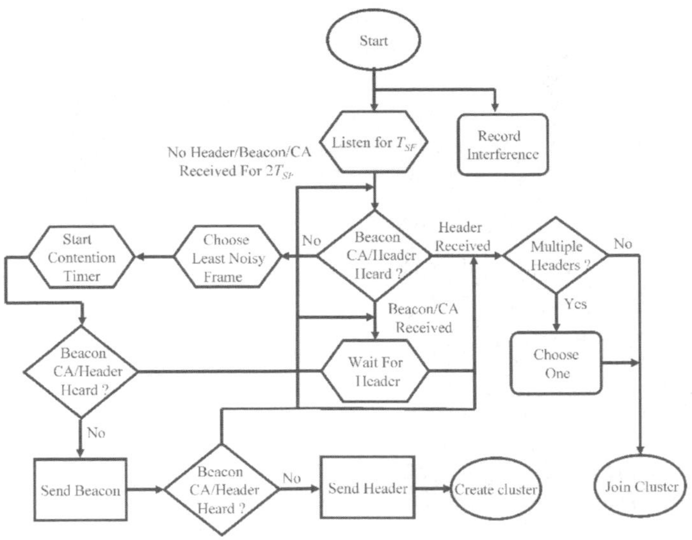


*Figure 7.4.* MH-TRACE cluster creation flow chart.

### 7.2.4 Cluster Formation and Maintenance

At the initial startup stage, a node listens to the medium to detect any ongoing transmissions for the duration of one superframe time, $T_{SF}$, to create its interference table for each frame within the superframe. If there is already a clusterhead in its receive range, the node starts its normal operation. If more than one beacon is heard, the node that sent the beacon with higher received power is chosen as the clusterhead (*i.e.*, the closest clusterhead is chosen). If no beacon is detected, then the node chooses the least noisy frame, picks a random time within that frame to transmit its own beacon signal, and begins to listen to the channel until its contention timer expires. If a beacon is heard in this period, then the node just stops its timer and starts normal operation. Otherwise, when the timer expires, the node sends a beacon and assumes the clusterhead position. In case there is a beacon collision, none of the colliding nodes will know it, but the other nodes hear the collision, so the initial startup continues. All the previously collided nodes, and the nodes that could not detect the collision(s) because of capture, will learn of the collisions with the first successful header transmission. Cluster creation is presented as a flow chart in Figure 7.4.

Each clusterhead continuously records the interference level of each frame by listening to the beacon transmission and CA transmission slots, which are at the beginning of each frame. Since only the clusterheads are allowed to transmit in these slots, it is possible for each clusterhead to measure the received power level from other clusterheads and know the approximate distances to other

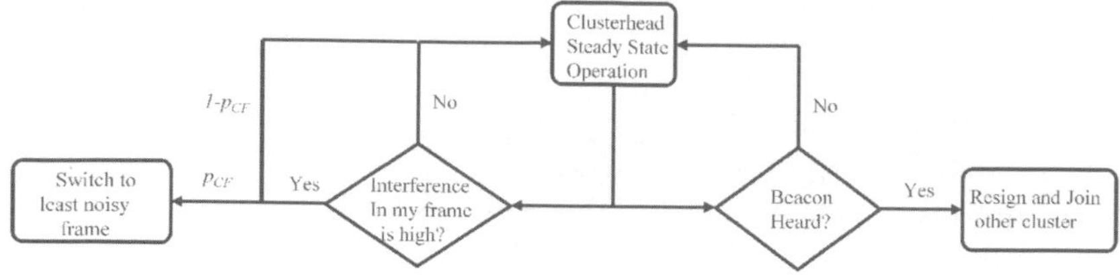


*Figure 7.5.* MH-TRACE cluster creation flow chart.

clusterheads in the carrier sense range. A clusterhead can record the interference level of each frame by listening to the beacon slot, but the beacon slot becomes useless for a clusterhead's own frame, because it is transmitting its own beacon. A CA packet, which is transmitted with a probability $p_{CA}$, is used to determine the interference level of the co-frame clusters. If this probability is set to 0.5, then each clusterhead records the interference level in its frame, on the average, at $4T_{SF}$ time.

A clusterhead keeps its frame and continues to operate in its steady state mode unless another clusterhead enters in its receive range. When two clusterheads enter in each other's receive range, the one who receives the other's beacon first resigns directly. A clusterhead leaves a frame with high interference (*e.g.*, two clusterheads enter each other's interference range but not receive range) and moves to a low interference frame with probability $p_{CF}$. The reason for adding such randomness is to avoid the simultaneous and unstable frame switching of co-frame clusters, which are the interference source for each other. If $p_{CF}$ is set to 0.5, then the probability that only one of the two co-frame clusterheads switches to a new frame becomes 0.67. Cluster maintenance is presented as a flow chart in Figure 7.5.

If a node does not receive a beacon packet from its clusterhead for $2T_{SF}$ time, either because of mobility of the node or the clusterhead or the failure of the clusterhead, then it enters the initial startup procedure.

### 7.2.5 Dynamic Clusterhead Selection

The spatial traffic density in the network is a statistical distribution created by the temporal characteristics of the voice sources and the mobility pattern. Therefore, the network traffic distribution is not perfectly uniform, and traffic at a specific portion of the network may be temporarily higher than the rest of the network. Thus, some clusters have fewer channel allocation requests than they can support, which results in underutilization of the resources, and some clusters have higher demand than they can support, which results in call blocking.

Many nodes in the network are in the transmit range of more than one clusterhead, and the default action for these nodes is to choose to request channel access from the closest clusterhead. For these nodes, if all the data slots in the closest cluster are in use and another cluster in range has available data slots, they can contend for channel access from the further clusterhead with unused data slots rather than the one that is closer but does not have available data slots. Note that the available data slot information of the previous superframe is included in the Beacon packet. Figure 7.6 shows a snapshot of a portion of the network structure, where nodes A-G are clusterheads with transmission ranges represented by the circles around them and node X is an ordinary node with its receive range represented by the shaded disk. Node X has three clusterheads (E, F, and G) in its receive range. The closest clusterhead is G, but if G does not have available data slots, then node X can choose to request channel access from E or F depending on the availability of the data slots in these clusters. By incorporating this dynamic channel allocation scheme into MH-TRACE, one more degree of freedom is added to the network dynamics, which enables efficient utilization of the bandwidth and reduces the adverse affects of clustering.

### 7.2.6 Listening Cluster Creation

Nodes listen to the IS slots of each frame, and based on the information gathered from the IS slots they determine which data transmissions in that particular frame to receive. Each node knows the transmitting nodes in its receive range in advance through IS packets sent by them, even if the node is not in the receive range of the clusterheads of those nodes and cannot receive the transmission schedule directly. For example, node X in Figure 7.6, can receive data from nodes that are members of seven different clusters, and four of these clusterheads are not in the receive range of node X. This shows the flexibility of the MH-TRACE architecture.

Advantages of the listening cluster are threefold: (*i*) each node needs to be awake only in the data slots that are occupied and sleeps in the rest of the data slots, (*ii*) all the data collisions are known in advance and energy dissipation for listening to collisions is avoided, because if the (small) IS packets have collided than the corresponding (large) data packets will also collide, and (*iii*) a framework for data discrimination is created.

If data discrimination is utilized, then each node creates its listening cluster, which has a maximum of $N_{MAX}$ members, by choosing the closest nodes based on the proximity information obtained from the received power from the transmissions in the IS slots (other data discrimination criteria can also be used).

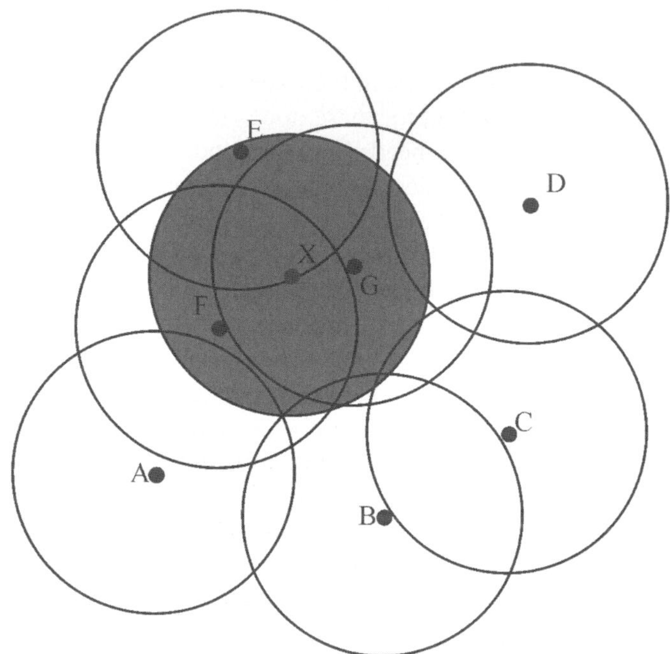


*Figure 7.6.* Network partitioning into clusters. Nodes A-G are clusterhead nodes, and the circles around them show their transmission radii. Node X is an ordinary node with its reception range shown with the shaded disk.

## 7.3 Simulations and Analysis

To test the performance of MH-TRACE and to compare it with other MAC protocols, we ran simulations using the ns-2 network simulator [128]. We simulated conversational voice coded at 32 Kbps, which corresponds to one voice packet per superframe. The channel rate is set to 2 Mbps. We used a perfect channel without any loss or error models. All the simulations are run with various numbers of nodes ranging from 50 to 200, moving within a 1 km by 1 km area for 100 seconds. We utilized the voice source model described in Section 6.3.2, which classifies speech into "spurts" and "gaps".

We used the propagation and energy models described in Section 2.2.1 and Section 5.1.1, respectively. We used the random way-point mobility model (see Section 2.4) to create mobility scenarios within a 1 km by 1 km area. Node speeds are chosen from a uniform random distribution between 0.0 m/s and 5.0 m/s (the average pace of a marathon runner) with zero pause time. For application scenarios confined to a 1 $km^2$ area, it is not practical to use high speed mobility patterns that are beyond pedestrian mobility (*i.e.*, vehicle mobility).

*Table 7.2.* Superframe parameters.

| Number of frames per superframe, $N_F$ | Number of data slots, $N_D$ | Number of contention slots, $N_C$ | Superframe time, $T_{SF}$, (ms) |
|---|---|---|---|
| 4 | 12 | 15 | 24.976 |
| 5 | 10 | 7 | 25.060 |
| 6 | 8 | 9 | 24.984 |
| 7 | 7 | 6 | 25.172 |
| 8 | 6 | 6 | 24.992 |

Beacon, CA, contention, and IS packets are all 4 bytes. The header packet has a variable length of 4-18 bytes, consisting of 4 bytes of packet header and 2 bytes of data for each node to be scheduled. Data packets are 104 bytes long, consisting of 4 bytes of packet header and 100 bytes of data. Each packet includes a 3-bit packet type field, an 8-bit source ID, an 8-bit preamble, and an 8-bit Cyclic Redundancy Check (CRC). Beacon and header packets also include a 4-bit number that specifies the number of slots currently in use, and IS packets include an end-of-stream bit. Each slot or sub-slot includes 16 $\mu s$ of Inter-Frame Space (IFS) to account for switching and round-trip time.

The simulations are repeated with the same parameters five times, and the data points in the figures are the average of the ensemble and the errorbars are the standard deviation of the ensemble. Acronyms, descriptions and values of the parameters used in the simulations are presented in Table 7.1.

### 7.3.1 Optimizing MH-TRACE Parameters

We investigated the effects of the number of frames, $N_F$, within the superframe on different aspects of the network operation through theoretical analysis and through simulations in a 100 node network, which is dense enough, yet not too dense, to represent a general case. Table 7.2 shows the system settings for different $N_F$. These settings are adjusted to keep the superframe time, TSF, as close as possible to the voice packet generation period, TVP, which is 25 ms.

Figure 7.7 (a) shows the total number of clusterheads throughout the simulation time as a function of $N_F$. This is a measure of the clusterhead lifetime and cluster structure stability. The number of clusterheads is high for $N_F = 4$ 58.2 $\pm$ 19.3), and it reduces with increasing $N_F$, reaching 31.0 $\pm$ 3.7 at $N_{F8}$. For simplicity, we are going to use $N_{F4}$ for $N_F = 4$. In $x \pm y$ notation, $x$ and $y$ are the mean and standard deviation of an ensemble, respectively. For lower $N_F$, the number of clusterheads is higher because of a higher number of collisions. Beacon packets of co-frame clusterheads collide at some regions of the network, and nodes in these areas cannot receive the beacon packets from

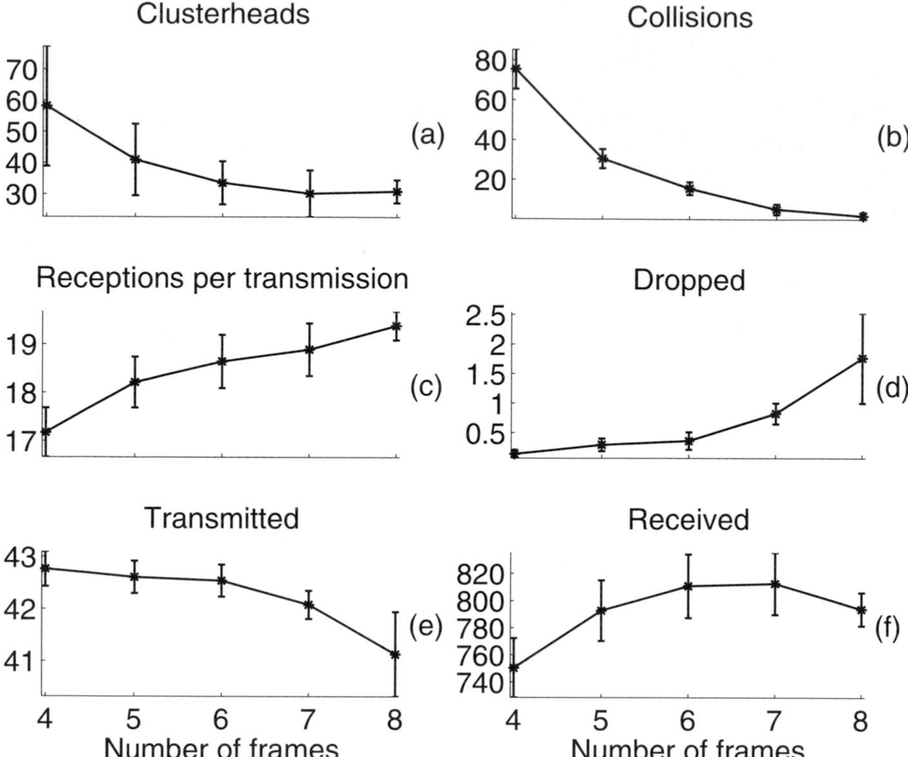


*Figure 7.7.* (a) Total number of clusterheads throughout the entire simulation time versus number of frames. (b) Average number of data packet collisions per superframe. (c) Average number of data packet receptions per transmission per superframe. (d) Average number of dropped data packets per superframe. (e) Average number of transmitted data packets per superframe. (f) Average number of received data packets per superframe.

either of the clusterheads, even though they are in the transmission range of the clusterheads. Thus, these exposed nodes enter startup to create their own clusters in this situation, which results in the resignation of existing clusterheads. The average number of clusterheads per superframe lies in a very narrow band (*i.e.*, $10.8 \pm 0.8$) for all $N_F$, which shows that the differences in total clusterhead numbers are due to short term fluctuations. This problem is alleviated almost completely for higher $N_F$, because for higher $N_F$ (*i.e.*, $N_{F7}$ and $N_{F8}$) co-frame clusterheads are far enough apart to avoid beacon collisions. However, due to node mobility, there is a limit on the average clusterhead lifetime, $35.5 \pm 6.7$ s, independent of $N_F$, because after some time depending on the speed and direction of the clusterheads, they will enter each other's transmission range and the one who receives the other's beacon first resigns.

Figure 7.7 (b) shows the number of data collisions per superframe versus $N_F$. Since all the clusterheads choose the least interference frame for transmission, it is obvious that the distance between the co-frame clusterheads is an increasing function of $N_F$. Therefore, the number of collisions decreases from $75.5 \pm 10.0$ at $N_{F4}$ to $2.0 \pm 1.7$ at $N_{F8}$.

Figure 7.7 (c) shows the number of collision-free receptions per transmission versus $N_F$, which is obtained by dividing the number of transmissions by the number of receptions. The approximate theoretical value of the average number of neighbors, $N_{neighbor}$, of a node in the network can be obtained by multiplying the coverage area with node density, which is given by

$$N_{neighbor} = \frac{\pi D_{Tr}^2 N_N}{A_{network}} \tag{7.1}$$

where $A_{network}$ is the total network area, which is $10^6$ $m^2$, and the coverage area of a node is a disk with the transmission range, $D_{Tr} = 250$ m, as its radius. Using these values, $N_{neighbor}$ is obtained as 19.63 for $N_N$ (total number of nodes in the network) equal to 100. If there were no collisions, then the average number of receptions per transmission would be equal to $N_{neighbor}$. For example, if we had a fully connected single-hop network with a single transmitting node, then the number of receptions per transmission would be equal to the number of neighbors of the transmitting node. As shown in Figure 7.7 (c), the number of receptions per transmission converges asymptotically to the theoretical value ($N_{neighbor}$) with increasing $N_F$, starting at $17.2 \pm 0.5$ at $N_{F4}$ and reaching $19.4 \pm 0.3$ at $N_{F8}$. Deviations from the theoretical value are due to collisions, because collisions prevent nodes in the transmission range from receiving the transmitted packets, especially at lower number of frames.

Figure 7.7 (d) shows the average number of dropped packets per superframe versus $N_F$. Since the total number of clusters and cluster coverage are independent of $N_F$ and the number of data slots per cluster, $N_D$, is inversely proportional with $N_F$, the total bandwidth available is less for high $N_F$, which explains the increasing trend in dropped packets with increasing $N_F$.

Figure 7.7 (e) shows the average number of transmitted data packets per superframe, which is the difference between the number of generated data packets and dropped data packets. The average number of generated data packets, $N_G$, is a function of $N_N$ and the average spurt and gap durations ($m_s$ and $m_g$, respectively) and is given by

$$N_G = \frac{m_s}{m_s + m_g} N_N \tag{7.2}$$

The average number of generated data packets is 43 for a 100 node network.

Figure 7.7 (f) shows the average number of data packet receptions by the whole network per superframe, which is the total network throughput, versus the number of frames. The number of receptions is at it lowest, $750.4 \pm 21.8$, at $N_{F4}$, it reaches a maximum, $812.6 \pm 22.9$, at $N_{F7}$, and again drops to $793.8 \pm 12.3$ at $N_{F8}$. The relatively lower number of receptions at lowest (*i.e.*, $N_{F4}$) and highest (*i.e.*, $N_{F8}$) number of frames is due to the higher number of collisions and higher number of packet drops, respectively.

Systematic variations in various metrics in Figure 7.7 (a) - (f) are due to two primary mechanisms that are balancing the aggregate network throughput as a function of $N_F$, which are very similar to the spatial reuse and co-channel interference concepts in cellular systems [27]. The first is the packet loss due to collisions and the second is the throughput loss due to dropped packets. We denote the function that gives the throughput loss due to collisions in terms of packets per frame as a function of $N_F$ as $f_{coll}$. The function that gives the throughput loss due to the dropped packets is denoted as $f_{drop}$, which is related to the average number of dropped packets per superframe, $N_{drop}$, through the equation

$$f_{drop} = N_{drop} N_{neighbor} \tag{7.3}$$

$N_{drop}$ is multiplied by $N_{neighbor}$ because each transmitted packet increases throughput by the number of one-hop neighbors of the transmitting node. In other words, $f_{coll}$ is the number of packet receptions that could not be realized due to collisions and $f_{drop}$ is the number of packet receptions that could not be realized due to the non-transmission of the packets that are dropped at the transmitters. The function that represents the total packet loss due to collisions and packet drops as a function of $N_F$, denoted as $f_{loss}$, is the sum of $f_{drop}$ and $f_{coll}$.

Figure 7.8 shows $f_{loss}$, $f_{coll}$, and $f_{drop}$ obtained from simulations and theory as functions of $N_F$. Both logical reasoning and simulation results show that $f_{drop}$ is a monotonic increasing function of $N_F$ and $f_{coll}$ is a monotonic decreasing function of $N_F$, respectively. $f_{loss}$, which is the summation of these two, is not monotonic. The reason that $f_{drop}$ is an increasing function is that for higher $N_F$, the number of available data slots per unit area is smaller and nodes experience more contention. On the other hand, for smaller $N_F$, separation between the co-frame clusters is less and the number of collisions is higher, which explains the decreasing characteristics of $f_{coll}$. The exact mathematical modeling of $f_{drop}$ and $f_{coll}$ is a challenging task, which necessitates joint analysis of temporal and spatial interactions of various random variables. Therefore we created a semi-analytical model for the characterization of these functions through curve fitting to the simulation data.

The general form of $f_{drop}$ is:

$$f_{drop}(N_F) = C_{drop} e^{K_{drop} N_F} \tag{7.4}$$

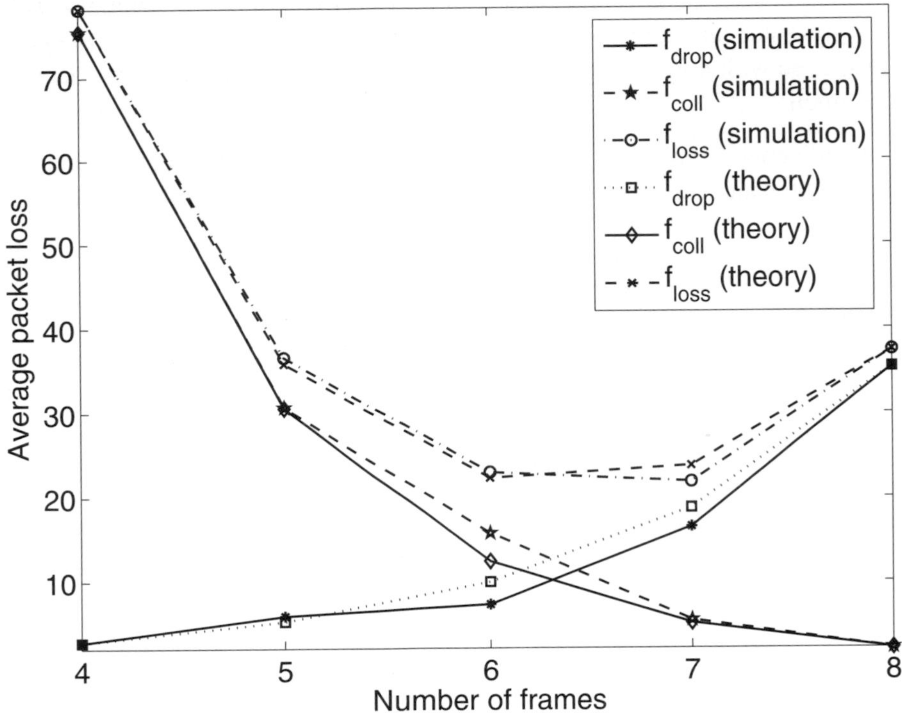


*Figure 7.8.* Average packet loss per superframe versus number of frames.

The constants in the equation, $C_{drop}$ and $K_{drop}$, are found to be 0.2 and 0.6, respectively. The general form of $f_{coll}$ is:

$$f_{coll}(N_F) = C_{coll}e^{K_{coll}N_F} \tag{7.5}$$

where $C_{coll} = 2816.3$ and $K_{coll} = 0.9$. The total throughput loss is:

$$f_{loss}(N_F) = f_{drop}(N_F) + f_{coll}(N_F) \tag{7.6}$$

Minimizing the total packet loss maximizes aggregate throughput. Based on the analysis above, we find that $N_{F7}$ provides minimum packet loss (23 packets per superframe) and maximum aggregate throughput (812 packets per superframe). Simulation results presented in Figure 7.8 also show that the optimal value of $N_F$ is 7. Although these simulation results are for a specific node density (*i.e.*, 100 nodes / 1 $km^2$), simulations with different node densities (*i.e.*, 50 nodes / 1 $km^2$ and 200 nodes / 1 $km^2$), which are not shown, also verify that the optimal $N_F$ value is seven. We will use $N_{F7}$ for the rest of the simulations. Note that the reason for choosing exponential functions was that they were the best fit to the data.

Nevertheless, the difference between the maximum and minimum throughput, presented in Figure 7.8 (f), is small (*i.e.*, less than 8.0% difference). Thus, even with non-optimal $N_F$, MH-TRACE performance does not deteriorate much.

More generalized and extensive investigation of a modified cluster creation and maintenance algorithm for MH-TRACE is presented in Appendix B.

### 7.3.2 Dynamic Clusterhead Selection

We investigated three clusterhead selection methods. The first method is to choose the closest clusterhead, denoted as $C_{HC}$, the second method is to choose the closest clusterhead with available data slots, denoted as $C_{HCA}$, and the third method is to choose the clusterhead with the maximum number of available data slots regardless of proximity, denoted as $C_{HA}$. Since the available data slot information of the previous superframe is included in the Beacon packet and proximity can be obtained by using the received power strength of Beacon packets, both availability and proximity information are already present at each node.

Figure 7.9 (a) shows the average number of aggregate received packets per frame versus $N_N$, the number of nodes, for $C_{HC}$, $C_{HCA}$, and $C_{HA}$. Throughput obtained with both $C_{HCA}$ and $C_{HA}$ is higher than that of $C_{HC}$, and the difference increases with increasing $N_N$. $C_{HCA}$ and $C_{HA}$ have very close values for all $N_N$, but $C_{HCA}$ is slightly better than $C_{HA}$ for $N_N = 200$. The difference between $C_{HCA}$ and $C_{HA}$ is due to the fact that $C_{HA}$ is more vulnerable to collisions than $C_{HCA}$ (see Figure 7.9 (b)), because it does not use the proximity information unless all the clusterheads in a node's receive range have the same number of available data slots. Simulation results show that decreasing the number of dropped packets is more important than avoiding collisions (see Figure 7.9 (c)), because $C_{HC}$, which has fewer collisions but a higher number of dropped packets, has lower throughput than $C_{HCA}$ and $C_{HA}$, which have more collisions but a lower number of dropped packets. Although the node distribution is fairly uniform, especially for higher node densities, due to the statistical time dependence of the traffic, there are temporal non-uniformities in the spatial distribution of the data traffic. The difference between the clusterhead selection algorithms arises because of this fact. Since $C_{HC}$ does not take these non-uniformities into account, it cannot compensate for such non-idealities. On the other hand, both $C_{HCA}$ and $C_{HA}$ can deal better with this problem. It seems that $C_{HCA}$ and $C_{HA}$ have very similar characteristics, with $C_{HCA}$ having a slightly better throughput for denser networks. Therefore, we opted to use $C_{HCA}$ as the clusterhead selection algorithm for the simulation results presented in this chapter.

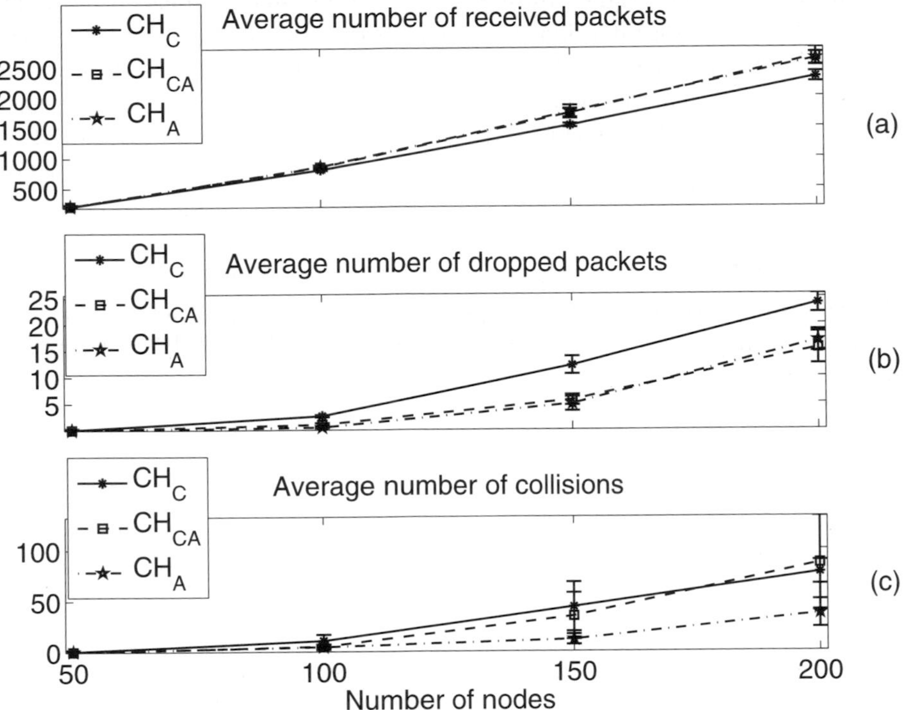


*Figure 7.9.* Comparison of clusterhead selection methods. (a) Average number of received packets per superframe versus number of nodes. (b) Average number of dropped data packets per superframe. (c) Average number of data packet collisions per superframe.

### 7.3.3 IEEE 802.11 and SMAC Simulation Models

We obtained quantitative comparisons of MH-TRACE, IEEE 802.11 and SMAC for various metrics. There are two main reasons to compare MH-TRACE with IEEE 802.11 and SMAC: (*i*) Both of these protocols are well known by the wireless community, and almost all researchers compare their algorithms with IEEE 802.11, making it possible to compare MH-TRACE with any other protocol by just comparing the performance relative to IEEE 802.11, and (*ii*) SMAC is the most prominent example of a truly distributed energy aware MAC protocol.

We modified the original SMAC protocol [22], which is presented in Section 5.1.2, to compare it with MH-TRACE on a fair basis. Actually, we take the basic design philosophy of SMAC, which is letting the nodes sleep periodically to save energy, and modified IEEE 802.11 to create the modified SMAC. Since we assumed global synchronization for MH-TRACE, we also assumed global synchronization for SMAC, so there are no synchronization packets and overhead in the modified SMAC. We tested several sleep/active ratios, and the optimal schedule (*i.e.*, highest throughput) for SMAC is a 25 ms sleep and 25 ms active cycle. Since the node density and packet generation rate in our framework is much higher than the cases tested in [22], several modifications are needed to optimize SMAC, like randomization of the contention start time after the sleep period for the packets that arrived during the sleep period and were stored for transmission in the awake period. If all the nodes with stored packets begin contention at the beginning of the active period, almost all the packets would collide, because it is not possible to comply with such high medium access demand at once for the underlying IEEE 802.11 contention resolution algorithm. We reduced the overhead for IEEE 802.11 and SMAC broadcast data packets to four bytes in our simulations to compare MH-TRACE with IEEE 802.11 and SMAC on a fair basis; therefore, data packets are 104 bytes for IEEE 802.11, SMAC and MH-TRACE.

### 7.3.4 Throughput

Figure 7.10 shows the average number of packet receptions per node per superframe versus the number of nodes for MH-TRACE, IEEE 802.11, SMAC, MH-TRACE with maximum listening cluster size of 5 (*i.e.*, $lc - 5$), MH-TRACE $lc - 10$, and the theoretical maximum throughput, which is obtained by multiplying the number of generated packets with the average number of neighbors, $N_{neighbor}$. The theoretical maximum is actually an upper bound, which can be achieved by eliminating packet drops and collisions.

For $N_N = 50$, throughput is very close for all cases and equal to $4.0 \pm 0.5$ packets/node/superframe, because at this node density there is not much contention for channel access and there is a large margin to be exploited to avoid packet drops (see Figure 7.11 (a)) and collisions (see Figure 7.11 (b)). MH-TRACE is closest to the theoretical maximum at all node densities, but it is also lower than the theoretical maximum throughput starting with $N_N = 100$, primarily due to packet drops. Referring to Figure 7.10, at $N_N = 200$, the theoretical maximum throughput, 17.4 packets/node/superframe, is 31% larger than MH-TRACE throughput, $13.3 \pm 0.7$ packets/node/superframe.

MH-TRACE $lc-5$ throughput converges to 5 packets/node/superframe starting with $N_N = 100$, because with lower node density the number of transmissions in a one-hop neighborhood of the nodes frequently drops below 5, so the average number of receptions cannot reach 5. For the same reason MH-TRACE $lc - 10$ throughput converges to 10 packets/node/superframe starting with $N_N = 150$.

The throughput of IEEE 802.11 is lower than MH-TRACE for $N_N > 50$, with an 86% difference at $N_N = 200$. Furthermore, IEEE 802.11 throughput starts to decrease for $N_N > 150$ ($7.9 \pm 0.2$ packets/node/superframe), which marks the limit of the stable operation in broadcasting for IEEE 802.11. For broadcast traffic, IEEE 802.11 does not use the standard four-way handshake

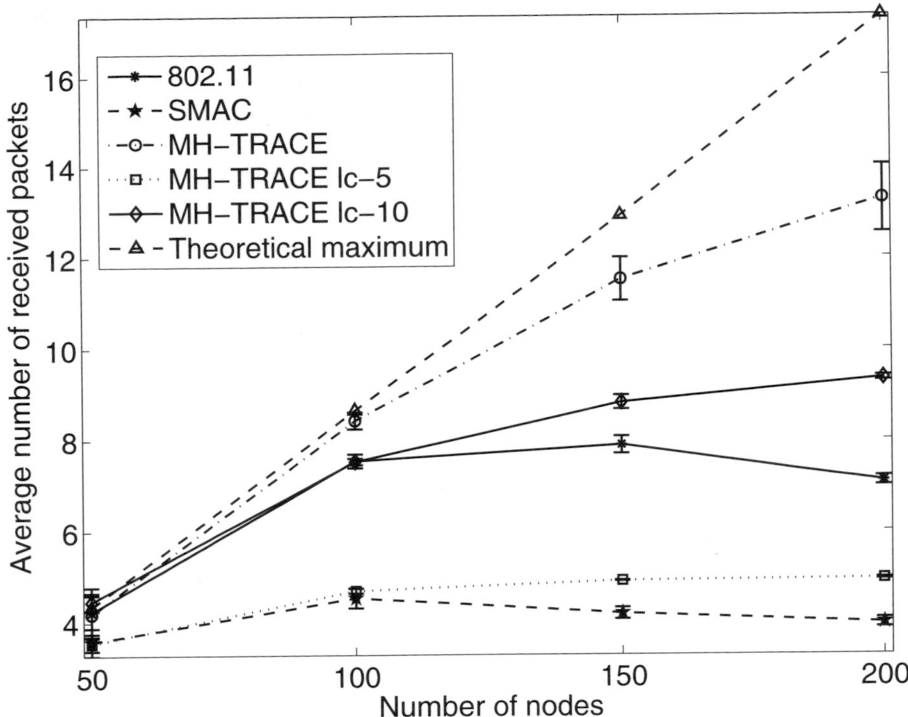


*Figure 7.10.* Average number of received packets per node per superframe versus number of nodes.

mechanism; instead, only the data packet is transmitted, since no feedback can be obtained from the other nodes, and binary exponential backoff (BEB) is not employed for broadcast traffic [20]. Thus IEEE 802.11 becomes Carrier Sense Multiple Access (CSMA) for broadcast traffic [19]. IEEE 802.11's contention resolution algorithm does a good job under low node densities, and its throughput is very close to the theoretical maximum. However, for dense networks (*i.e.*, $N_N > 50$) the lack of coordination significantly degrades the throughput of IEEE 802.11, eventually driving it to instability due to the unchecked increase in the number of collisions.

The throughput of SMAC at $N_N = 50$, $3.6 \pm 0.3$ packets/node/superframe, is close to that of IEEE 802.11, $4.2 \pm 0.6$ packets/node/superframe. However, at $N_N = 200$, the throughput of SMAC is lower than that of all the other protocols (56% of IEEE 802.11, 30% of MH-TRACE, and 23% of the theoretical maximum). SMAC reaches instability at $N_N = 100$, sooner than IEEE 802.11. The relatively low throughput of SMAC is due to the number of collisions, which is approximately 10 times that of MH-TRACE at $N_N = 200$, and packet drops, which is approximately double of that of MH-TRACE at $N_N = 200$.

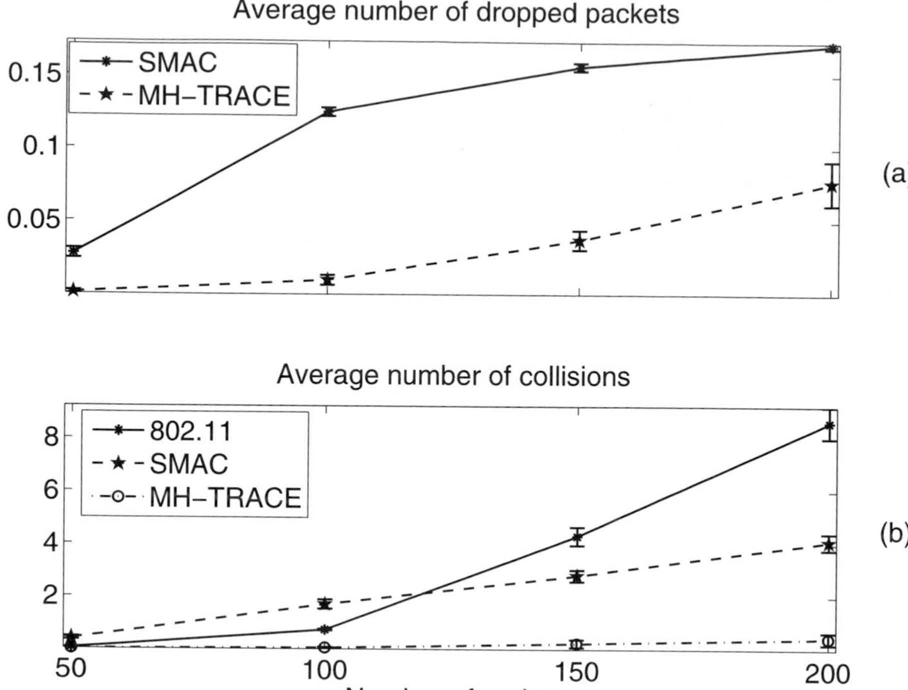


*Figure 7.11.* (a) Average number of dropped data packets per node per superframe versus number of nodes. (b) Average number of data collisions per node per superframe.

The basic design philosophy of SMAC, saving energy by reducing the active time, actually is equivalent to decreasing the bandwidth. In our simulations the sleep/active ratio is unity; thus half of the time is always unusable. However, the traffic handled in the awake period is more than half of the traffic (*i.e.*, more than 70% of the packets are transmitted, only 30% are dropped at $N_N = 100$). Thus the contention for medium access is more severe for SMAC than IEEE 802.11, which further degrades the already heavily loaded contention resolution algorithm of IEEE 802.11. The traffic adaptive sleep/active ratio adjustment mechanism of the original SMAC [22] cannot change the sleep/active ratio significantly due to the short packet transmission time, which is 0.416 ms.

### 7.3.5 Packet Delay

Figure 7.12 shows the average voice packet delay versus the number of nodes for MH-TRACE, IEEE 802.11, and SMAC. The average packet delay for MH-TRACE is an almost linear curve starting with $24.3 \pm 2.2$ ms at $N_N = 50$ and reaching $33.3 \pm 0.6$ ms at $N_N = 200$. Packet delay for IEEE 802.11 and SMAC also increases monotonically with increasing number of nodes, starting

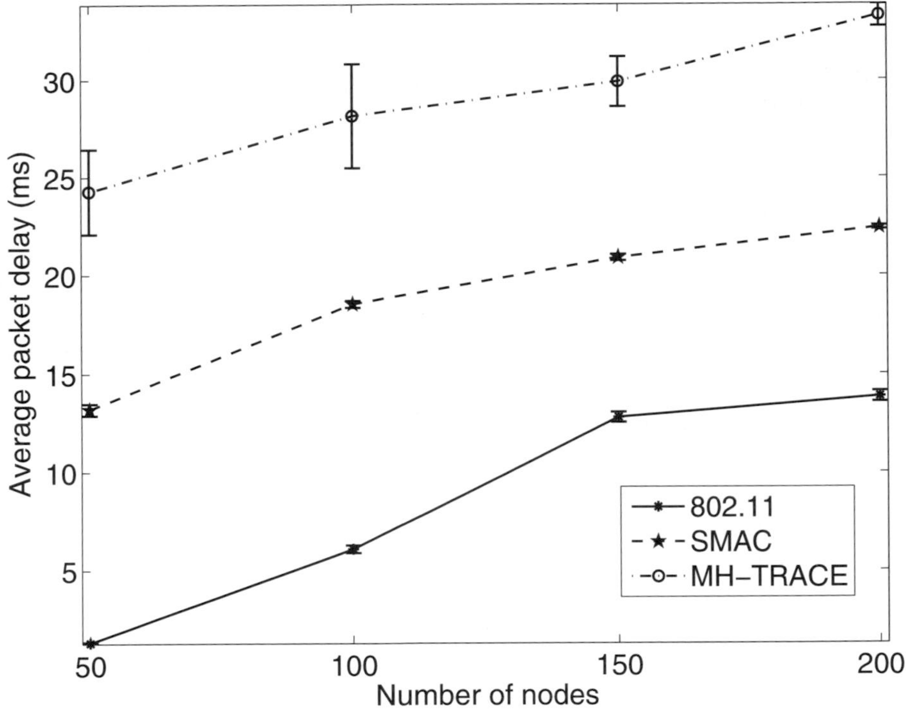


*Figure 7.12.* Average packet delay versus number of nodes.

with $1.3 \pm 0.04$ ms and $13.2 \pm 0.3$ ms at $N_N = 50$, and reaching $13.8 \pm 0.3$ ms and $22.4 \pm 0.1$ ms for IEEE 802.11 and SMAC, respectively.

Since IEEE 802.11 does not have an adaptive adjustment mechanism available for broadcasting, the backoff window is chosen to be an optimal value for a particular packet size and data traffic, which maximizes channel utilization and minimizes packet delay. Therefore, IEEE 802.11 cannot keep up with the varying data traffic. For example, for $N_N = 50$, the throughput obtained with IEEE 802.11 is as good as that of MH-TRACE and the delay is much lower, but for $N_N = 200$, IEEE 802.11 throughput is $54\%$ of the throughput obtained with MH-TRACE and the delay is still comparatively lower ($41\%$ of MH-TRACE packet delay). For data packets, lower delay is better, but for voice packets this is not always true. A voice packet with a 50 ms delay, the maximum packet delay allowed by the MAC layer after which the packets are dropped, and another voice packet with a 1.0 ms delay are equivalent from the application's point of view, which shows that QoS is an application dependent concept and should be considered in the design of all layers of the protocol stack. MH-TRACE exploits this feature of voice packets to tradeoff the packet delay for throughput and energy efficiency.

Packet delay in MH-TRACE is directly related with superframe time. Thus, it is possible to reduce the packet delay by shortening the superframe time. Superframe time can be shortened by: (*i*) keeping the number of frames within the superframe constant and reducing the number of data slots in each frame and (*ii*) keeping the number of data slots in each frame constant and reducing the number of frames within the superframe. However, any mismatch between the superframe time, $T_{SF}$, and voice packet generation period, $T_{VP}$, will create problems in the automatic renewal of channel access, because nodes that already gained channel access will not have a voice packet at each superframe. This problem can be alleviated by renewing the channel access in an interleaved fashion (*i.e.*, if the packet generation time is $N$ times the superframe time, then the channel access will be granted to each continuing voice stream at each $N$'th superframe). However, reducing the superframe time and incorporating additional control functionality will increase the system complexity and decrease the bandwidth used for data transmission due to increased overhead.

### 7.3.6 Energy Dissipation

Figure 7.13 shows the energy dissipation per node per superframe versus node density for IEEE 802.11, SMAC, MH-TRACE, MH-TRACE with no energy saving by staying active all the time (MH-TRACE-NES), MH-TRACE $lc-5$, MH-TRACE $lc-10$, and the theoretical minimum energy dissipation that is required to transmit and receive the same number of packets with MH-TRACE without any control packets, packet overhead, and energy dissipation for idle listening, collision reception, and carrier sensing. The dominant term in the theoretical minimum energy dissipation is due to packet receptions; therefore, the energy dissipation increases with the increase in throughput as a function of the number of nodes (see Figure 7.10).

Energy dissipation values of MH-TRACE at $N_N = 50$ and $N_N = 200$ are $1.04 \pm 0.04$ mJ and $2.32 \pm 0.04$ mJ, respectively, which are 73.4% ($0.44 \pm 0.09$ mJ) and 23.9% ($0.64 \pm 0.64$ mJ) higher than the theoretical minimum, respectively. The extra energy dissipation is mostly due to control packet transmission and reception and data packet overheads.

The difference between MH-TRACE and MH-TRACE-NES is $3.29 \pm 0.09$ mJ at $N_N = 50$ and $4.50 \pm 0.06$ mJ at $N_N = 200$. In other words, MH-TRACE energy dissipation is 24% and 34% of the energy dissipation of MH-TRACE-NES, without losing any information, which shows that it is possible to achieve significant energy savings without degrading system performance in the MH-TRACE framework. The extra energy dissipation is mostly due to idle listening for lower node densities, but for higher node densities carrier sensing also becomes important. Energy dissipation for receiving packets above the

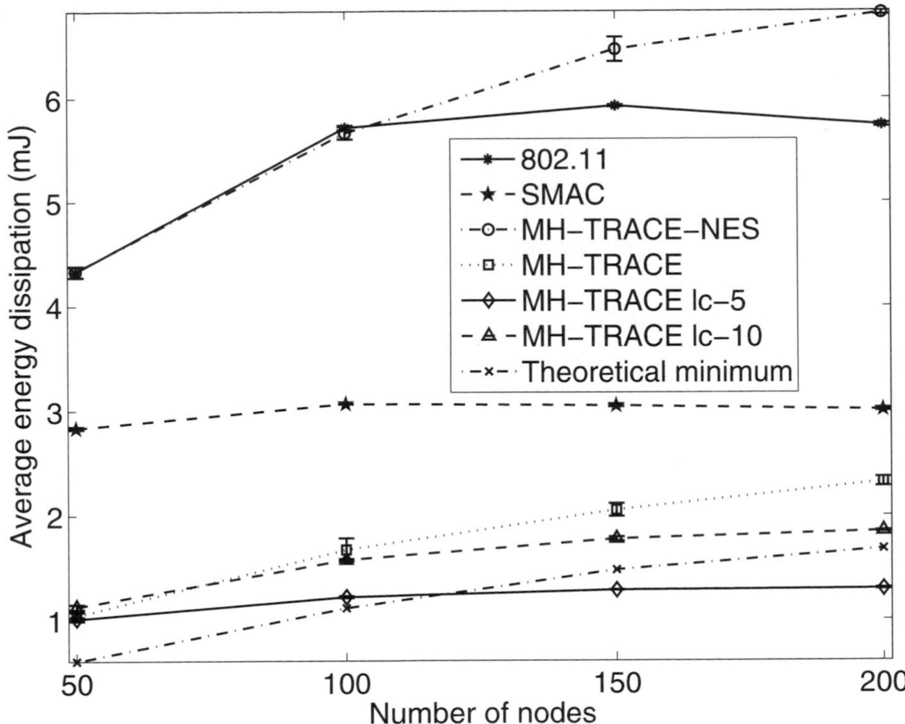


*Figure 7.13.* Average energy dissipation per node per superframe versus number of nodes.

reception threshold is the same as energy dissipation for receiving packets below the reception threshold but above the carrier sense threshold [107]. Performing carrier sense for beacon and CA packets is necessary for the clustering algorithm to run properly, but there is no point in performing carrier sense for the data packets-this is just a waste of energy for no gain.

MH-TRACE $lc-5$ and $lc-10$ dissipate almost the same energy as MH-TRACE at $N_N = 50$, because the average number of transmitting neighbors is not higher than the maximum listening cluster sizes at this node density. However, with increasing node density, energy savings by utilizing listening clusters becomes more evident. For example, at $N_N = 200$, the energy dissipation of regular MH-TRACE is 79% and 26% higher than that of MH-TRACE $lc-5$ and $lc-10$, respectively. This is because with higher node densities, the number of simultaneously transmitting nodes exceeds the maximum listening cluster sizes of 5 and 10 for $lc-5$ and $lc-10$, respectively.

Energy dissipation of IEEE 802.11 and MH-TRACE-NES are close for $N_N < 150$, because the number of transmissions and receptions (either successful or collided) are close to each other. However, starting with $N_N = 150$, which is the limit of stability for IEEE 802.11, IEEE 802.11 has lower energy dissipation than MH-TRACE-NES because the total number of collisions and successful receptions of IEEE 802.11 is lower than that of MH-TRACE in instability conditions. Note that the energy dissipation for a collision reception is half of the energy dissipation for two successful receptions. The energy dissipation of IEEE 802.11 is much higher than the energy dissipation of MH-TRACE for all node densities: the energy dissipation of MH-TRACE is 24% of that of IEEE 802.11 with the same throughput at $N_N = 50$, and at $N_N = 200$, the energy dissipation and throughput of MH-TRACE is 40.0% and 187% of those of IEEE 802.11, respectively.

SMAC energy dissipation stays in a narrow band, $2.96 \pm 0.11$ mJ, for all node densities, with a maximum of $3.07 \pm 0.003$ mJ at $N_N = 100$, which is the limit of stability, and a minimum of $2.84 \pm 0.01$ mJ at $N_N = 50$. When compared with MH-TRACE, SMAC dissipates 171 % and 30 % more energy at $N_N = 50$ and $N_N = 200$, respectively. Extra energy dissipation for lower node densities are mainly due to the idle listening and carrier sensing, and for higher node densities it is primarily due to collision reception. Energy dissipation of SMAC is 37% and 77% of that of IEEE 802.11 at $N_N = 50$ and $N_N = 200$, respectively. The energy savings of SMAC over IEEE 802.11 is due to the sleep period and fewer packet receptions due to packet drops, which results in degraded throughput and increased packet delay.

Energy savings of MH-TRACE are affected by many parameters including transmit, receive, idle, and sleep powers, node density, and maximum listening cluster size. The amount of energy saved is lower if the transmit power is much higher than the receive power, the sleep power is close to the idle power, and the idle power is much less than the receive power. On the other hand, the amount of energy saved is higher if the transmit and receive powers are close, idle power is close to the receive power, and sleep power is much less than receive power. These parameters are dependent on the radio electronics, and radios with both of the above specifications exist. In our simulations we used an actual radio model, which is midway between the above two extremes. If the node density, maximum listening cluster size, and ratio of transmitting nodes to total nodes are high, then the amount of energy saved is lower, because all the radios need to be on for extended durations to receive all the data packets. If the listening cluster size is low, then independent of the node density and ratio of the transmitting nodes, the amount of energy savings is higher.

It has been shown that there is an optimum transmit radius beyond which single-hop transmission is less energy-efficient than multi-hop transmissions [120, 121, 108]. By following the methodology in [120], we found that the maximum energy-efficient transmit range for our radio and propagation models is 326.0 m. Thus, our transmission range, which is 250.0 m, is in the energy-efficient range.

## 7.4 Discussion

The number of packets, packet sizes, and interframe space, which is the time to account for the guard times between the slots and time required to switch from one mode to another (*i.e.*, receive, transmit, sleep, idle), are very important factors in protocol performance. If the interframe space is long (*i.e.*, on the order of milliseconds - satellite systems or slow radio electronics), then the best thing to do is to reduce the number of packets, because even if the packet size is very small, the time slot required for this transmission is long. If the interframe time is small but the overhead in the data packets is high when compared to the payload, then again it is better to use a minimum number of packets to both save energy and increase throughput. Therefore, MH-TRACE operates as an energy-efficient and high throughput protocol if the interframe space is not extremely long and the overhead in the packets is not too high.

MH-TRACE is very sensitive to clock mis-synchronization, and the maximum tolerance is one $IFS$ time, which is 16 $\mu s$. Any clock mis-synchronization beyond $IFS$ would destroy the interference monitoring and clustering mechanisms of MH-TRACE. Network-wide synchronization can be achieved by using commercial GPS receivers, which are reported to have 200 ns accuracy [131] and are capable of operating indoors [132]. However, using a GPS receiver will increase the cost and energy dissipation of the radios. Network-wide synchronization can also be achieved by running a synchronization algorithm, which does not need GPS. In [133], it is reported that their synchronization algorithm achieves a maximum difference of 3.68 $\mu s$ within a 4-hop neighborhood using off-the-shelf IEEE 802.11 cards without any external references. Actually, network-wide synchronization is also crucial in IEEE 802.11 [134] and Bluetooth [135] networks for Frequency Hopping Spread Spectrum (FHSS) and Direct Sequence Spread Spectrum (DSSS) operation.

The distribution of the nodes in the network also affects the performance of MH-TRACE. MH-TRACE is designed to operate properly in a network with uniform node density. Since no clusterheads can be in each other's transmission range, clusterheads are distributed uniformly, on average, throughout the whole network. Therefore, the best case for MH-TRACE is a network with uniform node distribution. A network with nodes concentrated in a very small area is the worst case for MH-TRACE, because there will be only a few clusterheads and only a small portion of the available bandwidth can be used. IEEE 802.11 also performs better in a uniform node distribution, which results in uniform contention throughout the network assuming the traffic generated by each node is statistically equivalent.

Both white noise and bursty noise are factors that degrade protocol performance. If the white noise level of the network is beyond the carrier sense threshold, then cluster creation and maintenance will be negatively affected from this factor. However, it is also true for IEEE 802.11 that if the noise level is beyond the carrier sense threshold, then the radios will always sense the medium busy and the protocol operation suffers. Bursty noise is a hardship that cannot be thwarted easily. If a high power burst comes during a packet transmission, even if the burst duration is less than the packet duration, most probably the whole packet becomes useless. MH-TRACE is more sensitive to bursty noise than IEEE 802.11, because in broadcasting there are only data packets in IEEE 802.11. On the other hand, there are more control packets than data packets in MH-TRACE. For example, if the schedule packet is corrupted than the whole frame becomes useless. However, the control packets are much shorter than the data packets, and it has been shown that the probability of packet loss is smaller for shorter packets [136]. In Chapter 8, we further investigate the effects of channel noise on MH-TRACE performance.

## 7.5 Summary

In this chapter we discuss our design and evaluation of MH-TRACE, which is a MAC protocol that combines advantageous features of fully centralized and fully distributed networks for energy-efficient real-time packet broadcasting in a multi-hop radio network. We introduce a novel clustering algorithm that dynamically organizes the network into 2-hop clusters. MH-TRACE clusters are just for coordinating channel access and minimizing interference; thus, ordinary nodes are not static members of any cluster. Time is organized into cyclic superframes, which consist of several time frames, to support reservation-based periodic channel access for real-time traffic. Each clusterhead chooses the frame with least interference based on its own measurements for the operation of its cluster. Energy dissipation for receiving unwanted or collided data packets or for waiting in idle mode is avoided through the use of information summarization (IS) packets sent prior to the data transmissions by the source nodes. Through the use of transmission schedules within each cluster, managed by the clusterheads, intra-cluster data collisions are completely eliminated and inter-cluster collisions are minimized. We investigated MH-TRACE through extensive simulations and theoretical analysis. Our results show that MH-TRACE outperforms existing distributed MAC protocols like IEEE 802.11 and Sensor MAC (SMAC), in terms of energy efficiency and throughput, approaching the theoretical maximum throughput and theoretical minimum energy dissipation.

MH-TRACE does not need a routing protocol for the local broadcasting scenarios we considered in this chapter. However, for network-wide broadcasting, a routing protocol, which might be designed as a separate layer or embedded into the MAC layer, is needed. Chapter 10 concentrates on extending MH-TRACE to network-wide voice broadcasting.

# Chapter 8

# EFFECTS OF CHANNEL ERRORS

## 8.1 Introduction

MAC protocols can be classified into two categories based on the collaboration level of the network in regulating the channel access: coordinated and non-coordinated. A coordinated MAC protocol operates with explicit coordination among the nodes and is generally associated with coordinators, channel access schedules and clusters. For example, Bluetooth is a coordinated MAC protocol, where channel access within a cluster (*i.e.*, piconet) is coordinated by a coordinator (*i.e.*, piconet Master) [2]. A non-coordinated MAC protocol, on the other hand, operates without any explicit coordination among the nodes in the network. For example, IEEE 802.11 is a non-coordinated MAC protocol when operating in the broadcast mode (*i.e.*, in broadcasting mode, IEEE 802.11 becomes plain CSMA without any handshaking) [8]. Note that IEEE 802.11 channel access in unicasting mode is a coordinated scheme (*i.e.*, the four way handshaking between the transmitter and receiver is a special case of a general explicit coordination scheme, such as [7, 10]).

Figure 8.1 illustrates the channel access mechanism for generic coordinated and non-coordinated MAC protocols. In the coordinated MAC protocol, node $N_0$ is the clusterhead (coordinator) for the portion of the network consisting of five nodes. Channel access is regulated through a schedule that is broadcast by the coordinator. Upon reception of the schedule, nodes transmit their data at their allocated time, and thus collisions among nodes within the same cluster are eliminated. Furthermore, a node can switch to a low-energy sleep mode during the slots where no transmissions are scheduled or scheduled transmissions are not of interest to a particular node. Time is organized into cyclic time frames, and the transmission schedule is dynamically updated at the beginning

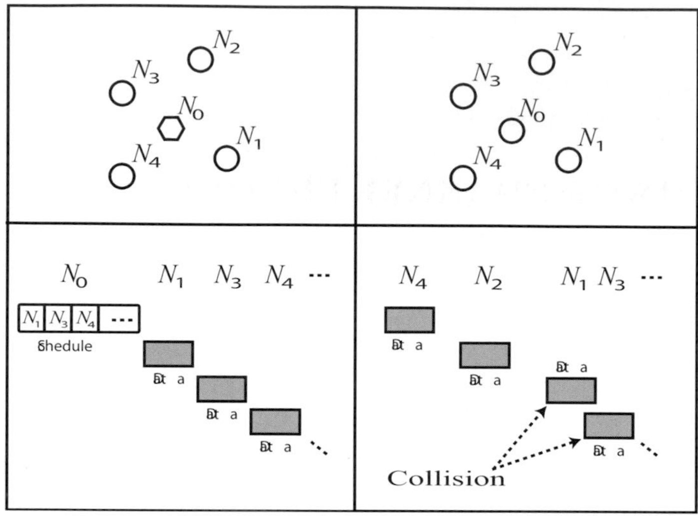


*Figure 8.1.* Illustration of coordinated and non-coordinated MAC protocols. The upper left and right panels show the node distributions for nodes $N_0$-$N_4$. The lower left panel shows the medium access for the coordinated scheme, where node $N_0$ is the coordinator and the channel access is regulated through a schedule transmitted by $N_0$. The lower right panel shows the channel access for the non-coordinated scheme (*e.g.*, CSMA). Overlapping data transmissions of $N_1$ and $N_3$ lead to a collision.

of each time frame. IEEE 802.15.3 [110] and MH-TRACE (see Chapter 7) are recent examples of such a coordinated MAC protocol. In the non-coordinated MAC protocol, each node determines its own transmission time based on feedback obtained through carrier sensing on the channel. Thus, conflicts in data transmission attempts (*i.e.*, collisions, capture) are unavoidable in the non-coordinated scheme. In addition, none of the nodes can switch to sleep mode because future data transmissions are not known beforehand due to the lack of a scheduling mechanism.

Both coordinated and non-coordinated MAC protocols have their advantages and disadvantages.

(i) One of the most important advantages of coordinated MAC protocols is their energy efficiency due to the availability of a schedule that lets nodes enter into sleep mode without deteriorating the overall system performance. Thus, the average energy dissipation of nodes in coordinated schemes is significantly lower than in non-coordinated schemes [35].

(ii) Collisions are mostly eliminated in coordinated MAC protocols, while frequent packet collisions are unavoidable in non-coordinated protocols, especially under heavy network conditions, which may draw the network into instability in extreme conditions [44].

(iii)The average packet delay using non-coordinated MAC protocols is lower than the average packet delay using coordinated MAC protocols under mild traffic loads. However, under heavy traffic loads, packet delay in non-coordinated protocols rises to very high levels [74].

(iv)Coordinated MAC protocols are more vulnerable to packet losses than non-coordinated MAC protocols due to their dependence on the reliable exchange of control packets, such as the schedule packet. Mobility, multi-path propagation, and channel noise are the main sources of errors that cause packet losses [137].

Energy efficiency has become one of the predominant platform requirements for battery powered mobile multimedia computing devices. Therefore, the new challenge is to provide QoS in an energy-efficient manner rather than focusing solely on QoS by ignoring the energy dissipation [138]. Consequently, there is a growing interest in energy-efficient design, mainly concentrating on MAC layer energy reduction techniques [35, 139, 140]. Most of the proposed solutions use TDMA as a MAC scheduling principle in order to utilize the benefits of having a schedule such as fairness, stability and energy efficiency by regulating the channel access, minimizing collisions and enabling power saving features, respectively.

The general trend in the evaluation of the performance of network protocols (*e.g.*, energy efficiency) is to ignore channel errors and assume a perfect channel [141]. Although the assumption of a perfect channel is reasonable in the initial design stage, further verification of a proposed protocol should consider error resilience.

In this chapter we investigate the performance of two MAC protocols, IEEE 802.11 and MH-TRACE (see Chapter 7), at different Bit Error Rate (BER) levels by providing an analytical model that is validated by ns-2 simulations [142–144]. IEEE 802.11 is a non-coordinated MAC protocol when it is used for broadcasting. MH-TRACE is an energy-efficient coordinated MAC protocol that relies on control packet exchanges for its operation. A comparative evaluation of IEEE 802.11 and MH-TRACE for real-time data broadcasting using a perfect channel (see Chapter 7) showed that the energy efficiency of MH-TRACE is much better than IEEE 802.11 [45]. However, due to the relatively complicated design of MH-TRACE, which relies on robust control packet exchange, the advantages of MH-TRACE over IEEE 802.11 are questionable under high BER levels.

The remainder of this chapter is organized as follows. Section 8.2 presents related work. In Section 8.3, we introduce an analytical model for the performance of MH-TRACE as a function of BER. Simulation results and analysis of both protocols under different BER levels are presented in Section 8.4. A summary of this chapter is presented in Section 8.5.

## 8.2 Related Work

Although performance analysis of ad-hoc networks has found some noticeable attention in the literature [145–153], there is little work done to explore the characteristics of different types of MAC protocols (*i.e.*, coordinated and non-coordinated) under varying channel noise. This chapter aims to answer the question of whether a coordinated MAC protocol preserves its superior performance, or whether its higher level of vulnerability due to the dependence on robustness of the control traffic makes it unstable under high BER levels.

In [35] MAC protocols are compared in terms of battery power consumption in order to emphasize the characteristics of an energy efficient MAC protocol. They concluded that reducing the number of contentions reduces the energy consumption. Moreover, reservation (*i.e.*, coordination and scheduling) is proposed as a better solution for messages with contiguous packets. However, energy efficiency under channel noise was not explored in this study.

More focused works investigating packet loss and error resilience can be found in [154] and [155]. These studies concentrated on identifying and characterizing possible sources of packet losses in ad-hoc wireless networks. Mobility and congestion are pointed out as the main reasons in mobile ad-hoc networks [154]. On the other hand, [155] takes collisions, error in radio transmission and Signal to Noise Ratio (SNR) variation into account as the main reasons for packet losses in mobile ad-hoc networks. They both provide simulation results to demonstrate the effects of each individual source of packet losses.

In [136] an adaptive frame length control approach, which is implemented at the MAC layer to compensate for rapidly varying channel conditions of wireless networks, is presented. They showed that by adjusting the frame length, there is much to be gained in terms of throughput, effective transmission range and transmitter power for wireless links. All of their outcomes stem from the assumption that the probability of error for a longer packet is higher than the probability of error for a shorter packet. Therefore, reducing the frame length when the channel conditions are worse will improve the throughput since the effective transmission range is increased. As a result of improved throughput, less energy is consumed due to the reduced number of retransmissions as we discussed earlier. However, in their analysis they did not consider the effects of channel noise on control packets.

However, none of the aforementioned studies provide sufficient insight on the error resilience, in general, and the vulnerability of control traffic to channel noise, in particular, and hence the performance evaluation of MAC protocols under various BER levels. Although the impact of channel errors on the control packets is crucial to the overall performance of coordinated MAC protocols, evaluation of coordinated MAC protocols under realistic channel errors has found little attention in the literature. In this chapter we investigate the effects of channel errors on the control traffic in a coordinated MAC protocol and determine the extent of performance deterioration.

## 8.3 Analytical Model

In this section we develop an analytical model to estimate the performance of MH-TRACE as a function of BER. However, our model essentially models a generic coordinated MAC protocol, thus, it is not necessarily specific to MH-TRACE and it can easily be extended to analyze any coordinated MAC protocol with little modification (*e.g.*, IEEE 802.15.3 [110] and EC-MAC [109]). For example, IEEE 802.15.3 has a similar channel access mechanism to MH-TRACE, where time is organized into cyclic superframes and channel access is granted through a control packet that includes the schedule (*i.e.*, a beacon packet). Therefore, modeling the performance of IEEE 802.15.3 will be essentially the same as our modeling of MH-TRACE. In our analysis we do not consider any error correction scheme, thus, if there is at least one bit error within a packet, then that packet is discarded. Random packet errors are independently introduced at the receivers.

If a protocol cannot maintain the desired level of performance, then its energy efficiency becomes meaningless. Thus, in order to achieve meaningful energy efficiency, it is absolutely necessary to make sure that a protocol does not deteriorate system performance while saving energy. MH-TRACE is a protocol designed primarily for energy efficiency, and simulations presented in Chapter 7 showed that under ideal channel conditions its energy efficiency is superior IEEE 802.11, a non-coordinated protocol. However, the question is whether MH-TRACE preserves its performance in the face of channel errors. In this section we seek the answer to this question through mathematical analysis supported with simulations.

### 8.3.1 Basic Model

To demonstrate our approach clearly and with a simple example, first we consider a fully-connected network with a small number of static nodes. The number of data slots in one superframe is high enough to support all of the nodes in the network (see Table 8.1). When there are no channel errors, all nodes should be able to transmit and receive without any packet drops or collisions. There will be only one clusterhead in the network due to the fact that there cannot be two clusterheads that can hear each other directly.

The number of data packets generated per node per second, ($DP_{node}$), is equal to the packet rate ($R_{packet}$) of MH-TRACE (*i.e.*, one packet per superframe

*Table 8.1.* Simulation parameters.

| Acronym | Description | Value |
|---|---|---|
| $T_{SF}$ | Superframe duration | 25.172 ms |
| $T_F$ | Frame duration | 3.596 ms |
| $N_F$ | Number of frames | 7 |
| $N_{DS}$ | Number of data slots per frame | 7 |
| $N_C$ | Number of contention slots per frame | 6 |
| $T_B$ | Beacon slot duration | 32 $\mu$s |
| $T_{CA}$ | CA slot duration | 32 $\mu$s |
| $T_C$ | Contention sub-slot duration | 32 $\mu$s |
| $T_H$ | Header slot duration | 92 $\mu$s |
| $T_{IS}$ | IS sub-slot duration | 32 $\mu$s |
| $T_D$ | Data slot duration | 432 $\mu$s |
| N/A | Data packet size | 104 B |
| N/A | Header packet size | 4-18 B |
| N/A | All other control packet size | 4 B |
| IFS | Inter-frame space | 16 s |
| $T_{drop}$ | Packet drop threshold | 50 ms |
| $T_{VF}$ | Voice packet generation period | 25.172 ms |
| $P_T$ | Transmit power | 0.6 W |
| $P_R$ | Receive power | 0.3 W |
| $P_I$ | Idle power | 0.1 W |
| $P_S$ | Sleep power | 0.0 W |
| $D_{Tr}$ | Transmission range | 250 m |
| $D_{CS}$ | Carrier Sense range | 507 m |

time $(1/T_{sf})$).

$$DP_{node} = R_{packet} = \frac{1}{T_{sf}} \qquad (8.1)$$

$DP_{node}$ represents the number of data packets generated by a single node in the network and can be regarded as the maximum number of packets a node can transmit given that it has full access to a perfect channel whenever it needs. However, a lossy channel will cause packet drops and therefore the throughput of the network will drop accordingly.

In Figure 8.2, the corresponding throughput losses due to dropped beacon, header and contention packets are given to illustrate the impact of the particular control packet on overall protocol performance. In these results only the specified control packets are lost due to channel errors and all the other packets are not affected [142].

These results are obtained from simulations with a six node fully connected static network to clearly observe the effects of packet losses. When there are

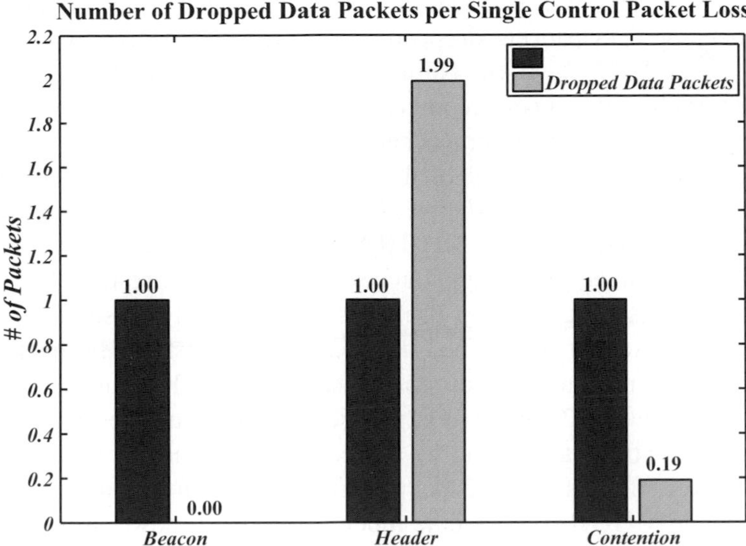


*Figure 8.2.* MH-TRACE performance degradation in terms of dropped data packets for beacon, header, and contention packet losses.

no channel errors, all nodes should be able to transmit and receive without any packet drops or collisions. There will be only one clusterhead in the network due to the fact that there cannot be two clusterheads that can hear each other directly. We utilized $1.0\%$, $3.0\%$, and $5.0\%$ packet error probabilities. Note that a $5.0\%$ packet error probability represents a harsh environment [137]. We used the ns-2 simulator to evaluate the system performance.

As can be seen from Figure 8.2, a lost header packet has the most impact on the performance of MH-TRACE. Loss of contention packets cause 10 times less impact on throughput than loss of header packets (0.19). Finally, for each beacon packet dropped, only 0.0015 data packets are dropped. Like beacon packet losses, losing other control packets (*e.g.*, IS, CA) do not significantly affect the throughput of the network. Thus, we conclude that the header and contention packets are the only control packets whose loss due to channel noise affect the network performance.

Therefore, we can write the equation for the *transmit* throughput of a single node (*i.e.*, transmit throughput per node per second $T_{node}$) in terms of the data packets dropped before transmission due to lost header packets ($DPL_H$) and contention ($DPL_C$) packets:

$$T_{node} = DP_{node} - DPL_H - DPL_C \qquad (8.2)$$

Both $(DPL_H)$ and $(DPL_C)$ can be written as the product of three parts. (*i*) Number of data packets dropped per header or contention packet loss $(DPL_{perH}$ or $DPL_{perC})$. (*ii*) Number of header or contention packets sent to a node or clusterhead per second $(HP_{node}$ or $CP_{node})$. (*iii*) Probability of dropping a header or contention packet $(P_H or P_C)$.

As contention packets are relatively short (4 bytes), they are less likely to be dropped than header packets (16 bytes for 6 broadcasting nodes). Furthermore, since the sources are CBR and MH-TRACE utilizes automatic channel access renewal, once a node gets channel access, it will not loose it and, thus, will not need to transmit contention packets for the rest of the simulation time. Moreover, the number of dropped data packets per lost header packet is 10 times larger than the number of dropped data packets per lost contention packet, as shown in Figure 8.2. Therefore, it is reasonable to assume that the effect of losing contention packets can be neglected. Furthermore, by ignoring the control packets other than the header packet, we focus on a more general model rather than an MH-TRACE specific model. Based on these assumptions, the transmit throughput per node per second becomes:

$$T_{node} = DP_{node} - DPL_H \tag{8.3}$$

$$T_{node} = \frac{1}{T_{sf}} - DPL_{perH} \times HP_{node} \times P_H. \tag{8.4}$$

In Equation (8.4), $DPL_{perH}$ is a constant (1.99) and $HP_{node}$ is equal to $DP_{node}$ since each node receives one header per super frame from its clusterhead. Finally $P_H$ depends on the length of the header packet $L_H$ and is calculated from the Bit Error Rate (BER) of the channel.

$$P_H = \{1 - (1 - BER)^{L_H}\}. \tag{8.5}$$

Therefore,

$$T_{node} = \frac{1}{T_{sf}} - 1.99 \times \frac{1}{T_{sf}} \times \{1 - (1 - BER)^{L_H}\} \tag{8.6}$$

$$T_{node} = \frac{1}{T_{sf}} \times [1 - 1.99 \times \{1 - (1 - BER)^{L_H}\}]. \tag{8.7}$$

In order to get the number of received packets per second in the network, we need to multiply the transmit throughput per node per second with the number of neighboring nodes $N - 1$ (note that all the nodes can hear each other in this network). Moreover, each data packet is received with a probability $P_D$, which is the probability that a data packet (with length $L_D = 104$ bytes) goes through the channel with no error at a given BER. Accordingly, the receive throughput per node per second $(T)$ becomes:

$$T = (N - 1) \times T_{node} \times (1 - BER)^{L_D}. \tag{8.8}$$

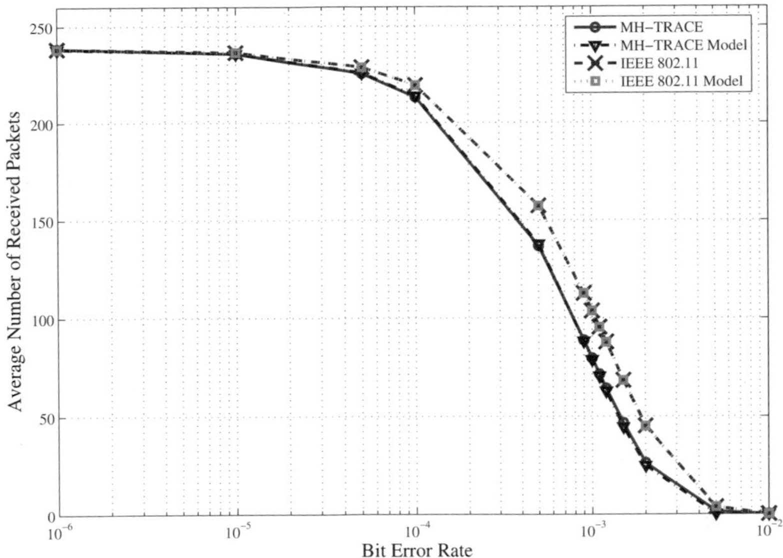


*Figure 8.3.* Average number of received packets per node per second versus bit error rate (BER).

Note that the receive throughput per node per second of IEEE 802.11 is simply equal to $\frac{N-1}{T_{sf}} \times (1 - BER)^{L_D}$ since in CSMA-type protocols such IEEE 802.11 in broadcasting mode, only data packets are sent through the lossy channel and the throughput is determined by the BER of the channel and length of a data packet.

We used the ns-2 simulator to validate the analytical model. The channel rate is set to 2 Mbps, and all nodes have a CBR data source with 32 Kbps data rate, which corresponds to one voice packet per superframe. The simulations are run for 1000 s and repeated with the same parameters five times.

In Figure 8.3, the analytical model for MH-TRACE and IEEE 802.11 are plotted against increasing BER. Also, the simulation results are included for both protocols to demonstrate the accuracy of the models. The throughput of MH-TRACE drops by almost 50% at a BER around $7 \times 10^{-4}$. On the other hand, IEEE 802.11 retains almost 55% of its initial throughput at the same BER (note that the initial throughputs of both protocols are the same). This difference can be translated into the fact that IEEE 802.11 performs 10% better than MH-TRACE, which experiences a worse performance degradation due to lost coordination packets [142].

These results show that the analytical model proposed to estimate the throughput of MH-TRACE is quite accurate. The model captures the fact that coordinated MAC protocols are more vulnerable than non-coordinated MAC protocols to channel noise due to their dependence on the robustness of the control traffic. In Figure 8.3, MH-TRACE experiences a steeper loss than IEEE 802.11 for BER values greater than $10^{-4}$, which is the point where header packet losses become the dominant factor in performance degradation. Our model captures this unique behavior of MH-TRACE. Therefore, our first conclusion is that the increased throughput loss occurs when a coordinated MAC protocol starts to lose its control packets. In our case, header packets are lost first since they are the longest control packet in MH-TRACE (see Table 8.1).

Before starting to derive a general model for MH-TRACE throughput, we want to mention that in our model, we treated the clusterhead as a regular node inside the network, but in reality, a clusterhead would not drop any data packets due to lost header packets since the clusterhead is the one generating the header packets. Therefore, our model slightly underestimates the throughput of MH-TRACE by treating the clusterhead as an ordinary node.

### 8.3.2 General Model

In this section, we consider a rectangular field ($L \times H$) in which a certain number of nodes ($N$), which have a communication radius ($r$), are randomly deployed. We use a statistical model of Voice Activity Detector (VAD) equipped voice source model that classifies speech into "*spurts*" and "*gaps*" (*i.e.*, gaps are the silent moments during a conversation). During gaps, no data packets are generated, and during spurts, data packets are generated at 32 Kbps data rate. Both spurts and gaps are exponentially distributed statistically independent random variables, with means $\eta_s = 1.0s$ and $\eta_g = 1.35s$, respectively [7]. The reason for using such a statistical voice source is that the transmission schedule will change frequently (*i.e.*, at the end of spurts nodes cease transmitting and their granted data slot will be taken away from them, and they will need to contend for channel access at the beginning of the next spurt), even in the absence of the channel errors, which is necessary to asses the system performance for a coordinated MAC protocol in the face of a dynamically changing transmission schedule.

Our approach to this more complex model will be basically the same as before. We begin by calculating the transmit throughput per node per second ($T_{node}$) when the channel is perfect. In addition to Equation (8.7), we need a term that captures the effect of the voice source model. This term can easily be represented with the ratio of spurts to the whole conversation ($\eta$). Therefore, we can write $T_{node}$ as in Equation (8.9).

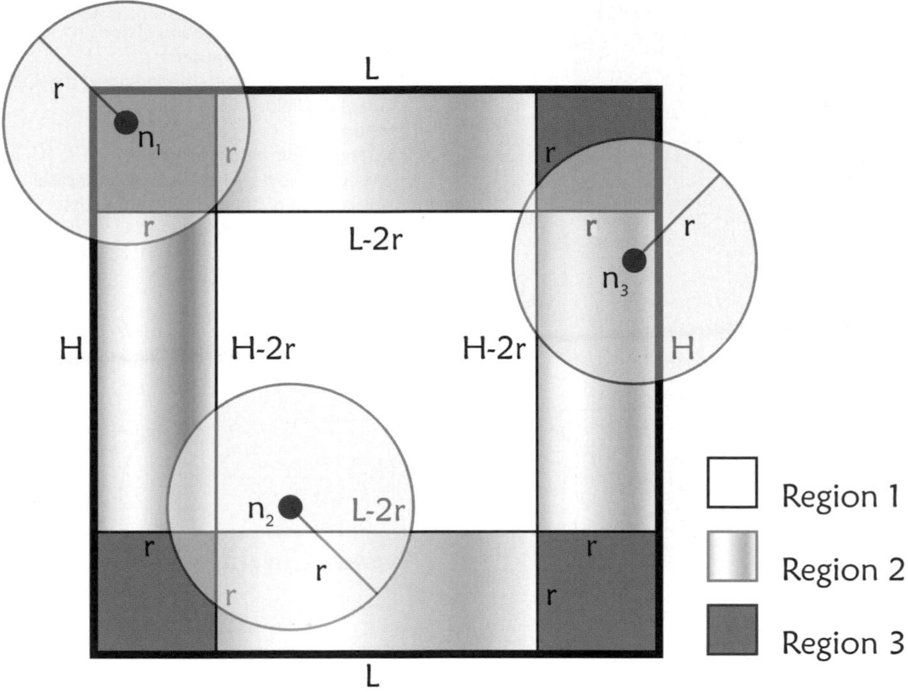


*Figure 8.4.* Rectangular field partitioned into three different regions.

$$
\begin{aligned}
T_{node} &= \frac{1}{T_{sf}}[1 - 1.99\{1 - (1 - BER)^{L_H}\}][\eta] \\
&= \frac{1}{T_{sf}}[1 - 1.99\{1 - (1 - BER)^{L_H}\}][\frac{\eta_s}{\eta_s + \eta_g}]
\end{aligned} \tag{8.9}
$$

After obtaining the expression for the transmit throughput per node per second, we have to find an expression for the average number of nodes within the communication range of a given node (*i.e.*, the average number of neighbors for a given node). In Figure 8.4, the rectangular field is partitioned into three different regions according to the coverage characteristic of a node in a particular region. For example, a node inside region 1 (*e.g.*, $n_2$) has its full coverage within the boundaries of the field. Therefore, any node inside region 1 utilizes 100% of its total coverage. Whereas nodes inside regions 2 and 3 (*e.g.*, $n_1$ and $n_3$) have a part of their coverage outside the field of interest and consequently the average percentage coverage for these nodes is less than 100%. Finding the percentage coverage for each region will lead us to the average number of neighbors.

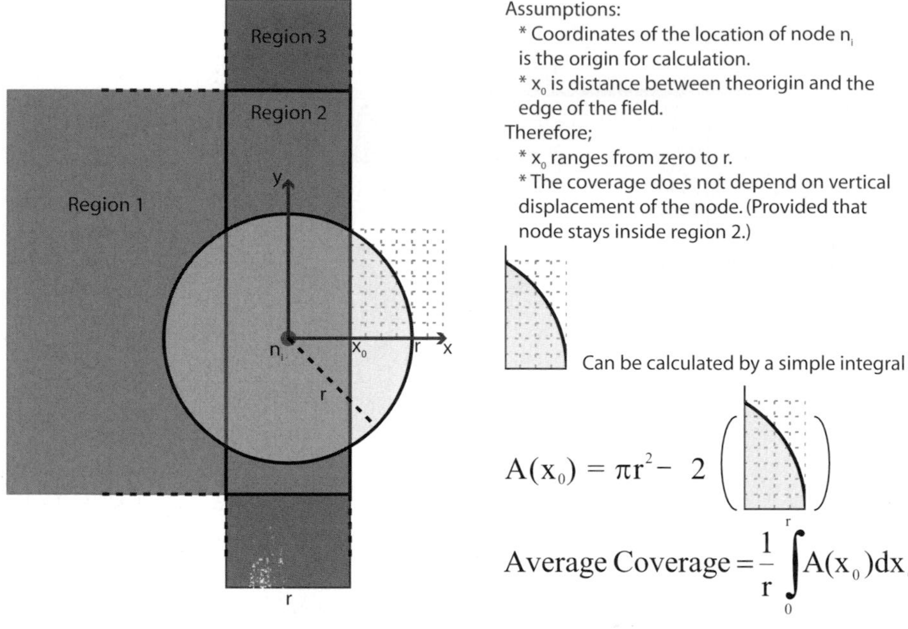

Region 3
Region 2
Region 1
y
x
$n_i$
$x_0$
r
r
r
Assumptions:
* Coordinates of the location of node $n_i$ is the origin for calculation.
* $x_0$ is distance between theorigin and the edge of the field.
Therefore;
* $x_0$ ranges from zero to r.
* The coverage does not depend on vertical displacement of the node. (Provided that node stays inside region 2.)
Can be calculated by a simple integral
$A(x_0) = \pi r^2 - 2$
Average Coverage $= \frac{1}{r}\int_0^r A(x_0)dx_0$


*Figure 8.5.* Calculation of the percentage coverage of a node inside region 2.

We start the derivation of the percentage with region 2. In Figure 8.5 the approach we used for obtaining the percentage is given. The area of the piece of circle shaded in Figure 8.5 can be expressed as follows:

$$\begin{aligned} I &= \int_{x_0}^{r} \sqrt{r^2 - x^2}dx \\ &= \frac{\pi}{4}r^2 - \frac{x_0}{2}\sqrt{r^2 - x_0^2} - \frac{r^2}{2}\arcsin(\frac{x_0}{r}). \end{aligned} \tag{8.10}$$

Thus, the average coverage for region 2 ($\alpha_2$) becomes,

$$\begin{aligned} \alpha_2 &= \frac{1}{r}\int_0^r A(x_0)dx_0 \\ &= \frac{1}{r}\int_0^r \left(\pi r^2 - 2I(x_0)\right) dx_0 \\ &= \pi r^2 - \frac{2}{3}r^2. \end{aligned} \tag{8.11}$$

After obtaining the average coverage as in Equation (8.11), we can easily calculate the percentage coverage of region 2 ($\sigma_2$).

$$\sigma_2 = \frac{\alpha_2}{\pi r^2} \quad = 1 - \frac{2}{3\pi} \tag{8.12}$$

Next we derive the average coverage for region 3 ($\sigma_3$). The area in question is divided into three parts (see Figure 8.6). According to this partitioning we have $A = \pi r^2 - (A_1 + A_2 - A_3)$, which is the coverage for a node inside region 3. The integrals for $A_1$ and $A_2$ are the same as $I$ given in Equation (8.10) and can be expressed as $2I(x_0)$ and $2I(y_0)$, respectively.

$$
\begin{aligned}
A_3 &= \int_{x_0}^{\sqrt{r^2-y_0^2}} \left(\sqrt{r^2-x^2} - y_0\right) dx \\
&= -\frac{y_0\sqrt{r^2-y_0^2}}{2} + \frac{r^2 \arcsin(\frac{\sqrt{r^2-y_0^2}}{r})}{2} \\
&\quad - \frac{x_0\sqrt{r^2-x_0^2}}{2} - \frac{r^2 \arcsin(\frac{x_0}{r})}{2} + y_0 x_0.
\end{aligned}
\tag{8.13}
$$

After obtaining $A_3$, we can calculate the average coverage $\alpha_3$ by taking the average of $A$.

$$
\begin{aligned}
\alpha_3 &= \frac{1}{r^2}\int_0^r \int_0^r A(x_0, y_0)\, dx_0 dy_0 \\
&= \pi r^2 - \frac{29}{24} r^2.
\end{aligned}
\tag{8.14}
$$

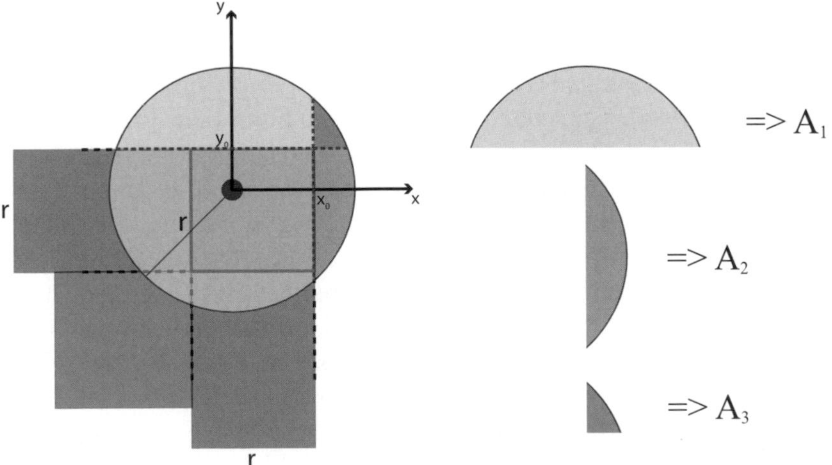


*Figure 8.6.* Calculation of the percentage coverage of a node inside region 3.

Thus, $\sigma_3$ becomes:

$$\sigma_3 = \frac{\alpha_3}{\pi r^2} = 1 - \frac{29}{24\pi} \tag{8.15}$$

This is the last percentage coverage we need to calculate the overall percentage coverage ($\sigma$), or the average number of nodes within the range of a given node inside the rectangular field. Below we give the resulting $\sigma$ in terms of the communication radius $r$, the length of the field $L$ and the height of the field $H$.

$$\sigma = \frac{\sigma_1(L-2r)(H-2r) + 2\sigma_2(H+L-4r)r + 4\sigma_3 r^2}{LH} \tag{8.16}$$

This expression can be used to calculate the average number of neighboring nodes ($N_N$) for a node inside of a rectangular field by multiplying $\sigma$ with $\pi r^2$ (*i.e.*, the coverage of a node with communication radius $r$) and the node density ($\frac{(N-1)}{LH}$). Note that there are $N-1$ nodes remaining that can be neighbors.

$$N_N = \frac{(N-1)\sigma\pi r^2}{LH} \tag{8.17}$$

Now, we can combine Equation (8.17) with Equation (8.9) to get the receive throughput per node per second, $T$.

$$T = N_N \times T_{node} \times (1 - BER)^{L_D} \tag{8.18}$$

According to our model, given that we have a constant simulation area and the same traffic model, throughput increases as the number of nodes in the network increases. In other words, the model suggests that throughput increases linearly with increasing node density. However, our previous work showed that throughput per node per second goes into saturation as the number of nodes in the network increases (see Figure 8.7). This trend is a result of packet collisions and drops emerging from mobility and increased contention for channel access [44]. According to this fact, we have to modify our initial throughput value (throughput when there is a perfect channel) in order to get a more accurate model for throughput. Since it is extremely challenging to model the dynamical behavior in Figure 8.7 analytically, the initial throughput values are calibrated according to feedback from simulation results.

## 8.4 Simulations and Analysis

### 8.4.1 Simulation Environment

To test the performance of MH-TRACE and IEEE 802.11 with increasing BER levels and to test the validity of our model, we ran simulations using the ns-2 network simulator [128]. We simulated conversational voice coded at 32 Kbps with VAD (see Section 8.3.2 and Section 6.3.2), which corresponds to

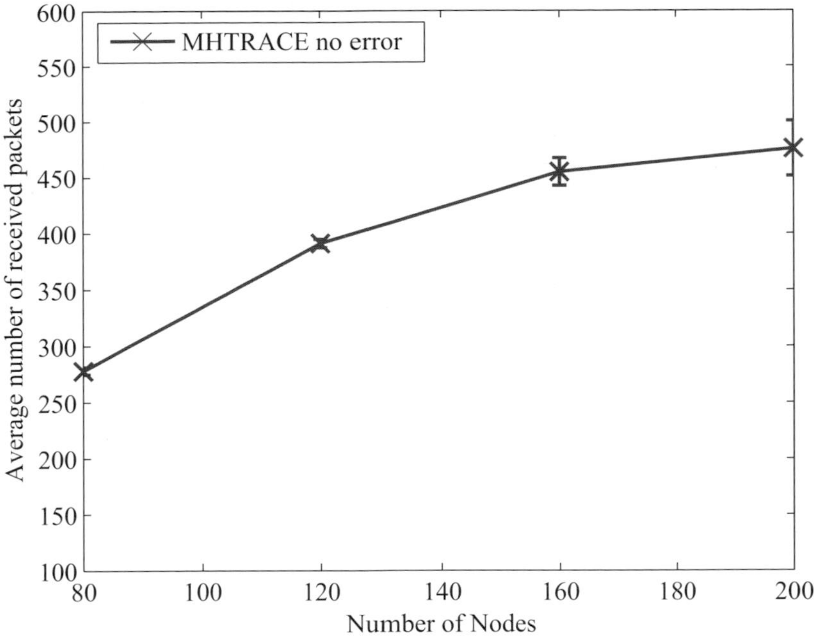


*Figure 8.7.* Average number of received packets per node per second versus number of nodes (mobile).

one voice packet per superframe. The channel rate is set to 2 Mbps and the standard IEEE 802.11 physical layer is employed for both protocols. All the simulations are run with 100 or 200 nodes, moving within a 1 km by 1 km area for 100 seconds according to the Random Way-Point (RWP) mobility model (see Section 2.4) with node speeds chosen from a uniform distribution between 0.0 m/s and 5.0 m/s (see Section 2.4). In this work we use 5.0 m/s, which is the average pace of a marathon runner, as our upper limit; however, we have observed that the performance of single-hop broadcasting in MH-TRACE does not change with increased mobility. Pause time is set to zero to avoid any non-moving nodes throughout the simulations. We used the propagation and energy models described in Section 2.2.1 and Section 5.1.1, respectively. The transport agent used in the simulations is UDP, which is a best effort service. The simulations are repeated with the same parameters six times, and the data points in the figures are the average of the ensemble. Acronyms, descriptions and values of the parameters used in the simulations are presented in Table 8.1 and Table 8.2.

Beacon, CA, contention, and IS packets are all 4 bytes. The header packet has a variable length of 4-18 bytes, consisting of 4 bytes of packet header and 2 bytes of data for each node to be scheduled. Data packets are 104 bytes long, consisting of 4 bytes of packet header and 100 bytes of data. Each slot or sub-slot includes 16 $\mu s$ of Inter-Frame Space (IFS) to account for switching and round-trip time.

In this study, we want to evaluate the performance of the MAC protocols; thus, the scenario we employ is single-hop data broadcasting, which does not require a routing protocol on top of the MAC protocol. Furthermore, in single-hop broadcasting, the overall performance (*e.g.*, energy dissipation, QoS, *etc.*) is directly determined by the performance of the MAC protocol.

*Table 8.2.* Simulation setup.

| Parameter | Value |
|---|---|
| Number of Nodes | 100 & 200 |
| Simulation Area | 1000m x 1000m |
| Simulation Time | 200s |
| Number of Repetitions | 6 |

## 8.4.2 Throughput

Figure 8.8 and Figure 8.9 present the throughput of MH-TRACE and IEEE 802.11 obtained from analytical models and simulations as functions of BER with 100 nodes and 200 nodes, respectively. Throughput is defined as the average number of received bit error free data packets per node per second. The analytical model developed in Section 8.3 (Equation (8.18)) for MH-TRACE is in very good agreement with the simulation results presented in the figures. The model for IEEE 802.11 is obtained by using the probability of successful data packet transmission ($(1 - BER)^{L_D}$) and the initial throughput value, which also closely follows the simulation results for IEEE 802.11.

The difference between the initial throughput values of MH-TRACE, where the BER rate is too low to affect the throughput (*i.e.*, BER $< 10^{-4}$), for the 100-node network (see Figure 8.8) and the 200-node network (see Figure 8.9) is due to the fact that both the number of transmissions and receptions are directly proportional to the total number of nodes in the network; thus, when the number of nodes is doubled, in ideal conditions, total throughput should be quadrupled. Hence, the throughput per node should be doubled. However, non-idealities, such as packet drops, keeps the throughput less than the ideal value. IEEE 802.11 throughput for the 200-node network is lower than the

100-node network throughput because of a very high collision rate. Note that while IEEE 802.11 collision resolution mechanism (*i.e.*, *p*-persistent CSMA in broadcasting) has a similar performance with MH-TRACE in the 100-node network, it becomes increasingly ineffective with the increasing node density (*i.e.*, IEEE 802.11 throughput is 60% of MH-TRACE throughput in the 200-node network).

There are two mechanisms that decrease the throughput of MH-TRACE with increasing BER: (*i*) with the increasing BER, more data packets are corrupted, which is also true for IEEE 802.11. Thus, the throughput decreases with increasing BER and (*ii*) the increase of the corrupted header packets results in unutilized data slots for MH-TRACE, whereas in IEEE 802.11 this is not a problem due to the lack of header packets. This situation creates an interesting tradeoff: while scheduling through header packets results in very high channel utilization in congested networks, it prevents nodes from channel utilization in high BER levels. However, when we examine the figures we see that MH-TRACE throughput is lower than IEEE 802.11 throughput only in low node density networks and only for extremely high BER levels (*i.e.*, the 100-node network and BER $\geq 10^{-3}$). Note that at BER $= 10^{-3}$ only 45% of the data packets are non-corrupted, which is not an acceptable operating condition. For all other situations, MH-TRACE throughput performance is better than IEEE 802.11. Furthermore, in the 200-node network MH-TRACE throughput

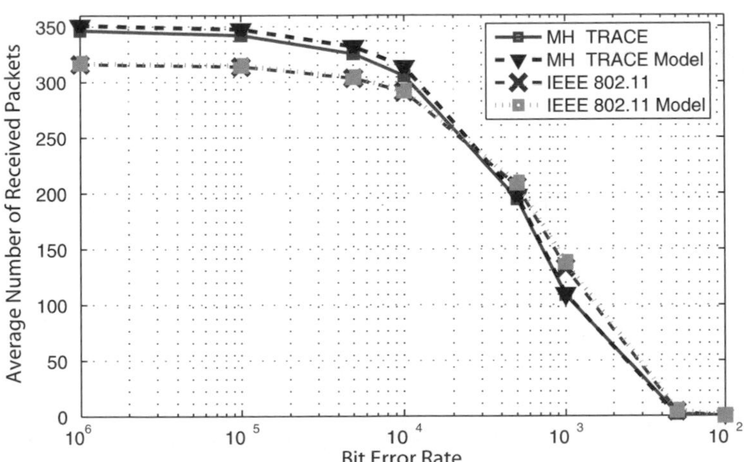


*Figure 8.8.* (100 nodes): Average number of received packets per node per second versus bit error rate (BER).

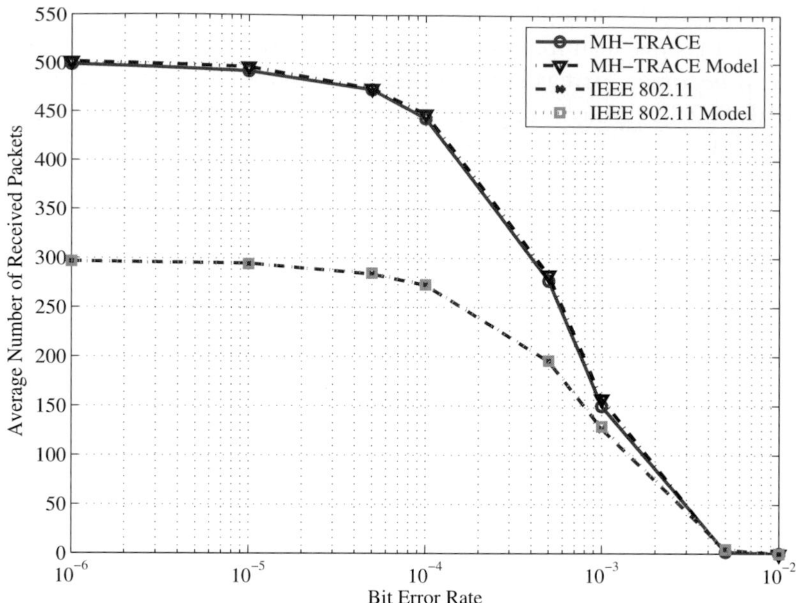


*Figure 8.9.* (200 nodes): Average number of received packets per node per second versus bit error rate (BER).

never drops below that of IEEE 802.11 throughput at any BER level. Thus, coordination through header packets is preferable over non-coordination regardless of the BER level of the network, especially in high congestion networks, from a throughput performance point of view.

### 8.4.3 Stability

Figure 8.10, presents the average clusterhead lifetime for the 100-node network and the 200-node network as a function of BER. Only the clusterheads that have a minimum lifetime of $10T_{SF}$ are counted in order to filter frequent clusterhead changes due to mobility and collisions so that only stable clusterheads are taken into consideration. Average clusterhead lifetime in the 100-node network is higher than the clusterhead lifetime in the 200-node network due the fact that the average number of clusterheads in a denser network is higher than the average number of clusterheads in a sparser network. This is because in sparse networks some areas are not covered by any clusterhead, and in fact, these areas are unpopulated by any node. On the other hand, in dense networks there are barely any uncovered areas. Thus, the total coverage of dense networks is higher, which can be made possible by higher number of clusterheads.

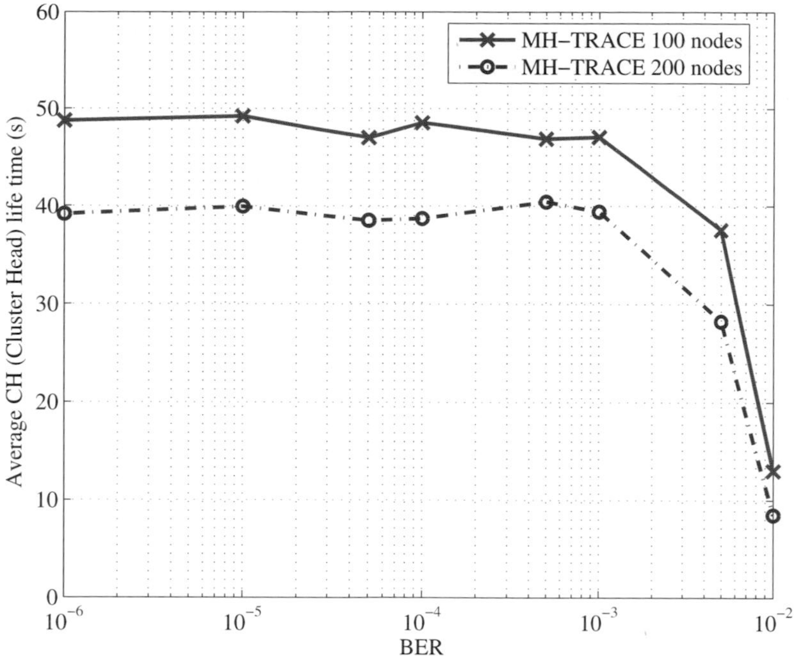


*Figure 8.10.* Average CH lifetime versus bit error rate (BER).

A higher number of clusterheads in the same network topology (*i.e.*, 1 km by 1 km network) results in less inter-clusterhead separation on the average, which increases the chance of one clusterhead moving into another's transmission range and resigning (*i.e.*, there cannot be any other clusterhead in the receive range of a clusterhead). Therefore, the average clusterhead lifetime in the 200-node network is lower than the average clusterhead lifetime in the 100-node network.

Clusterhead stability is not significantly affected by the BER level of the network for relatively low BER levels (*i.e.*, BER $\leq 10^{-3}$). This is because only 4 % of beacon packets are corrupted at BER $= 10^{-3}$, on the average. However, at BER $= 10^{-2}$, more than a quarter of the beacon packets are corrupted, which results in significantly shorter average clusterhead lifetime. Note that a node starts to contend for being a clusterhead if it does not receive a beacon packet for two consecutive superframes. Nevertheless, at BER $= 10^{-2}$, 99.98% of the data packets are corrupted. Thus, maintaining a cluster structure is not a meaningful consideration at such high BER levels.

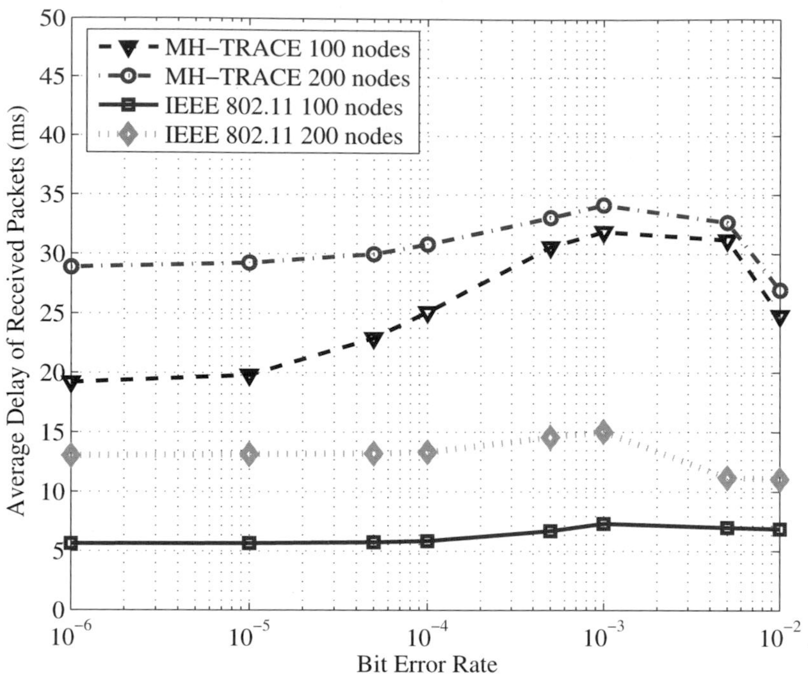


*Figure 8.11.* Average data packet delay versus bit error rate (BER).

### 8.4.4 Packet Delay

Figure 8.11 presents the average data packet delay for MH-TRACE and IEEE 802.11 as a function of BER. Data packets are dropped at the MAC layer if the data packet delay exceeds $T_{drop}$, which is 50 ms. MH-TRACE packet delay is higher than IEEE 802.11 packet delay at all BER levels in both the 100-node network and the 200-node network due to the fact that in MH-TRACE nodes can have channel access only once in a superframe time, whereas in IEEE 802.11 channel access is not restricted. MH-TRACE has comparatively higher packet delays as the BER level increases towards $10^{-3}$. The increase in the packet delay in MH-TRACE is mainly due to the header packet losses, as once a node looses a header packet, it loses several frame times before regaining channel access. IEEE 802.11 packet delay is almost constant. Packet delay is not very informative for BER levels higher than $10^{-3}$, because the throughput decreases to unacceptably low values. Average packet delay is higher in denser networks for both MH-TRACE and IEEE 802.11 due to the fact that higher channel utilization brings longer delays at the queue. The delay of MH-TRACE for

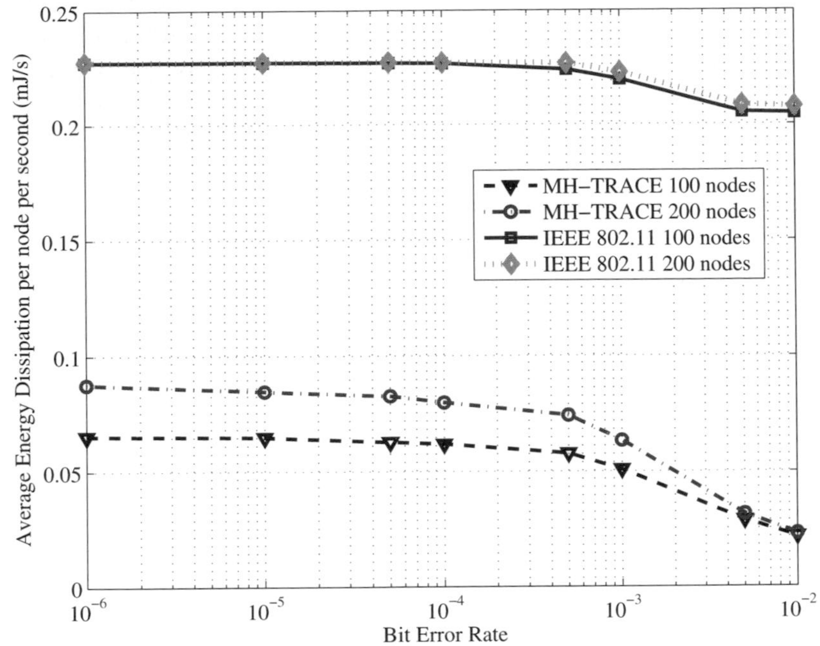


*Figure 8.12.* Average energy consumption per node per second versus versus bit error rate (BER).

both node densities tends to converge to similar values when BER $> 10^{-3}$. Although it is less obvious, the same behavior is also present in IEEE 802.11 case.

### 8.4.5 Energy Dissipation

One of the most important advantages of MH-TRACE over IEEE 802.11 is its better energy efficiency. Average energy dissipation per node per second for MH-TRACE and IEEE 802.11 with 100 and 200 nodes as a function of BER are presented in Figure 8.12. MH-TRACE energy dissipation under all BER levels and node densities is less than $40\%$ of the energy dissipation of IEEE 802.11.

IEEE 802.11 energy dissipation does not show any significant change with increasing BER due to the fact that the dominant energy dissipation terms in IEEE 802.11 are receive and carrier sensing and they are not significantly affected by the BER. This is because the energy dissipated for receiving a non-corrupted packet and a corrupted packet is the same. Furthermore, in carrier sensing only the presence of the carrier is important, which is not affected by the BER level of the network. Packet transmissions are also not related with BER level (*i.e.*, data packets are coming from the application layer and they are not routed). IEEE 802.11 energy dissipations in the 100-node network and the 200-node network are very close because the network is already in saturation conditions in the 100-node network (*i.e.*, full channel utilization) and this situation does not change in the 200-node network (*i.e.*, energy dissipated for a successful reception is the same with energy dissipated on a completely overlapping collision) from an energy dissipation point of view.

MH-TRACE energy dissipation is higher for the 200-node network than the 100-node network, because of the increase in channel utilization. Note that MH-TRACE is not utilizing all of the available bandwidth in the 100-node network (*i.e.*, a significant portion of the data slots are unused). The reason for the sharp decrease in energy dissipation of MH-TRACE for both node densities for BER $> 10^{-3}$ is that the nodes spend most of their time in sleep mode. Since a large portion of the header packets are corrupted, nodes cannot have channel access. Note that in MH-TRACE, a node is only awake if there is a scheduled data transmission. If the header packet is not received, then the corresponding node stays in the sleep mode for the whole frame time.

## 8.5 Summary

Energy efficiency of a MAC protocol is one of the most important performance metrics, especially in mobile ad hoc networks, where the energy sources are limited. Two key factors in achieving energy efficiency for a MAC protocol are coordination among the nodes and schedule-based channel access. In order to achieve a sufficient level of coordination among the nodes, and hence to achieve energy efficiency, the exchange of control information via control packets is vital. As such, coordinated MAC protocols (*e.g.*, MH-TRACE), which regulate channel access through scheduling, have been shown to achieve very high energy efficiencies when compared to non-coordinated MAC protocols (*e.g.*, IEEE 802.11), which do not employ scheduling. However, due to their increased vulnerability to channel errors, the performance of coordinated MAC protocols is affected more by the channel bit error rate (BER) than non-coordinated MAC protocols, which lack such control packets. In this chapter we investigated the energy efficiency and resilience against channel errors for coordinated and non-coordinated MAC protocols. Our results reveal that it is possible to achieve better system performance with coordinated MAC protocols, such as MH-TRACE, even in lossy channels, provided that the BER level is not extremely high.

# Chapter 9

# REAL-TIME DATA BROADCASTING

## 9.1 Introduction

In Chapter 7 we presented the MH-TRACE protocol, which is shown to have better energy efficiency and Quality of Service (QoS) support than the other Medium Access Control (MAC) protocols in single-hop voice broadcasting (*i.e.*, voice packets are not routed within the network). Although single-hop real-time data broadcasting has many applications (see Chapter 7), due to the limited radio range, single-hop broadcasting to all the nodes in the network is not possible in many ad hoc network scenarios, and thus multi-hop broadcasting is unavoidable. Although there are comparative studies on network-wide broadcasting algorithms and in particular on flooding, metrics such as QoS, energy dissipation, and the affects of the MAC layer on network-wide broadcasting have not received sufficient attention in the literature [12, 20, 156, 87]. Characterizing the effects of medium access control on the behavior of network-wide broadcasting is essential for designing high performance broadcasting architectures (network layer and MAC layer). In this chapter, flooding is utilized as the network layer broadcast algorithm due to its simplicity, which makes the role of the MAC layer more transparent and observable than more complicated broadcast algorithms. We investigate and quantify the QoS and energy dissipation characteristics of flooding when it is used for real-time data broadcasting for three different MAC protocols through extensive simulations and in-depth analysis [1, 46]. We believe that the results of this chapter provide a valuable contribution to the better understanding of QoS and energy efficiency for network-wide broadcasting.

The remainder of this chapter is organized as follows. Section 9.2 describes the broadcast architectures evaluated in this chapter. These broadcast architectures are IEEE 802.11-based flooding, Sensor MAC (SMAC)-based flooding, and Multi-Hop Time Reservation using Adaptive Control for Energy efficiency (MH-TRACE)-based flooding. The simulation environment is described in Section 9.3. Simulation results and analysis for the low traffic regime and high traffic regime are presented in Section 9.4 and Section 9.5, respectively. We present a summary of the simulations and analysis in Section 9.6.

## 9.2 Broadcast Architectures

In this chapter, we evaluate the QoS and energy dissipation characteristics of three flooding based network-wide broadcast architectures (IEEE 802.11-based flooding, SMAC-based flooding, and MH-TRACE-based flooding) within the (data rate, node density, network size/topology) parameter space. There are three main reasons for choosing these three MAC protocols to evaluate the performance of flooding: (*i*) the IEEE 802.11 standard is well known by the wireless community, and almost all researchers compare their algorithms with IEEE 802.11, making it possible to compare SMAC and MH-TRACE with any other protocol by just comparing the performance relative to IEEE 802.11, (*ii*) SMAC is a generic energy saving algorithm built on top of IEEE 802.11, and it represents a wide range of energy saving MAC protocols based on CSMA, and (*iii*) MH-TRACE is a MAC protocol specifically designed for energy-efficient single-hop real-time data dissemination. Furthermore, MH-TRACE is an example of a clustering based approach and a TDMA based channel access scheme. In this section, we provide brief descriptions of these architectures.

### 9.2.1 Flooding

Flooding is the simplest broadcasting algorithm, where each node rebroadcasts every packet it receives for the first time. Each node keeps track of the packets it received (*i.e.*, the source node ID and packet sequence number given by the source creates a unique global ID for each packet), and duplicate rebroadcasts are avoided. Furthermore, the sequence ID need not be more than the ratio of the packet drop threshold to the packet generation period in voice broadcasting (*i.e.*, 150 ms / 25 ms). Flooding is also a stateless algorithm, so the nodes do not need to create a routing framework (*e.g.*, routing tables, gateways, route caching, *etc.*). Despite some well known drawbacks, flooding is still used as a robust technique for information dissemination [157].

### 9.2.2 IEEE 802.11-based Flooding

In broadcasting mode, IEEE 802.11 uses $p$-persistent CSMA with a constant defer window length (*i.e.*, the default minimum defer period) [8]. When a node has a packet to broadcast, it picks a random defer time and starts to sense the channel (see Figure 9.1). When the channel is sensed idle, the defer timer counts down from the initially selected defer time at the end of each time slot.

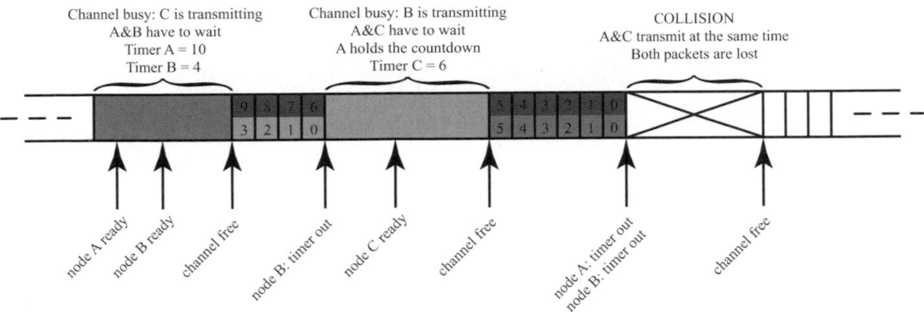


*Figure 9.1.* Illustration of the IEEE 802.11 medium access control mechanism in broadcasting.

When the channel is sensed busy, the defer timer is not decremented. Upon the expiration of the defer timer, the packet is broadcast.

However, when performing network-wide flooding, the contention resolution algorithm of IEEE 802.11 cannot successfully avoid collisions due to the high number of nodes contending for channel access concurrently. One method to avoid this problem is to spread out the packet transmissions at a higher level (*e.g.*, the network layer) by applying a random assessment delay chosen from a uniform distribution between $[0,\ T_{RAD}]$.

The IEEE 802.11 standard includes an energy saving mechanism when it is utilized in the infrastructure mode [8]. A mobile node that needs to save energy informs the base station of its entry to the energy saving mode, where it cannot receive data (*i.e.*, there is no way to communicate with this node until its sleep timer expires), and switches to the sleep mode. The base station buffers the packets from the network that are destined for the sleeping node. The base station periodically transmits a beacon packet that contains information about such buffered packets. When the sleeping node wakes up, it listens for the beacon from the base station, and upon hearing the beacon responds to the base station, which then forwards the packets that arrived during the sleep period. While this approach saves energy, it is not applicable in ad hoc mode, which we evaluate in this chapter.

### 9.2.3 SMAC-based Flooding

Many approaches have been proposed for reducing the energy dissipation of the IEEE 802.11 protocol [21, 158, 103, 159, 160]. Most of the work on energy-efficient MAC design based on Carrier Sense Multiple Access (CSMA) concentrates on unicast traffic. For example, Sensor MAC (SMAC) [160] is an energy-efficient MAC protocol designed specifically for sensor networks and built on top of IEEE 802.11. The designers make the observation that the main

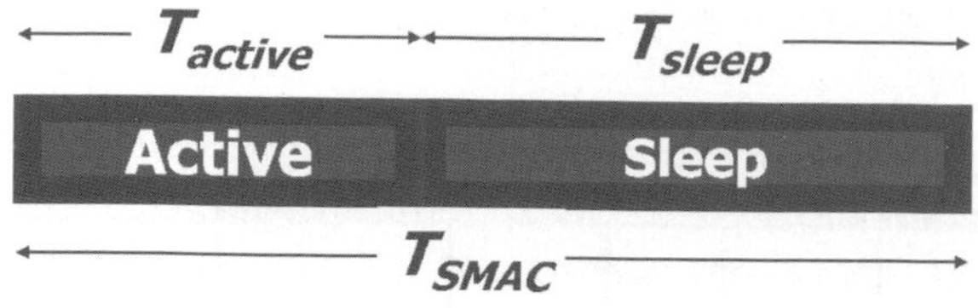


*Figure 9.2.* SMAC frame structure.

sources of energy inefficiency in IEEE 802.11 are idle listening and overhearing packets destined for other nodes. In SMAC, idle listening is reduced by periodically shutting the radios off. All the nodes in the network synchronize through synchronization packet broadcasts in a master-slave fashion to match their non-sleep periods. Furthermore, overhearing is avoided by entering the sleep mode after receiving the RTS and/or CTS packet until the NAV timer expires, which is matched to the duration of the data packet. It is shown that SMAC is much more energy-efficient than IEEE 802.11 in low data rate sensor network applications. However, it is not possible to employ the aforementioned energy saving mechanisms directly in broadcasting. Thus, we re-engineered the SMAC protocol for broadcasting as a representative for CSMA based energy saving protocols. Actually, we take the basic design philosophy of SMAC, which is letting the nodes sleep periodically to save energy, and modified IEEE 802.11 to create the modified SMAC.

In SMAC, time is organized into sleep/active time frames with duration $T_{SMAC}$, which repeat cyclically. Each frame is divided into two periods: (*i*) active period with duration $T_{active}$, where nodes can receive and transmit data, and (*ii*) sleep period with duration $T_{sleep}$, where nodes stay in a low energy sleep state (see Figure 9.2). The ratio of the sleep period in each sleep/active cycle, $R_{SMAC}$, is determined according to the QoS requirements of the application. Higher sleep/active ratios will result in higher energy savings at the expense of reduced effective bandwidth (*i.e.*, reduction of the actual usable time corresponds to an effective reduction of the bandwidth).

In SMAC, sleep/active mode switching is synchronized throughout the network (*i.e.*, we assume global synchronization, which is available through the Global Positioning System). In active mode, SMAC operation is similar to IEEE 802.11. However, if at the end of an active period a packet is not transmitted, then it is delayed until the sleep period ends, which increases the packet delay when compared to IEEE 802.11. Several modifications are needed to optimize SMAC, like randomization of the contention start time after the sleep period for the packets that could not be transmitted in the previous active period.

### 9.2.4 MH-TRACE-based Flooding

In Chapter 7 we present a detailed description of Multi-Hop Time Reservation Using Adaptive Control for Energy efficiency (MH-TRACE), which is a MAC protocol designed for energy-efficient real-time single-hop data broadcasting [45]. In this section we provide a brief description of MH-TRACE and the details of its integration with flooding. In MH-TRACE, the network is partitioned into overlapping clusters through a distributed algorithm (see Figure 7.1. Time is organized into cyclic constant duration superframes consisting of several frames. Each clusterhead chooses the least noisy frame to operate within and dynamically changes its frame according to the interference level of the dynamic network. Nodes gain channel access through a dynamically updated and monitored transmission schedule created by the clusterheads, which eliminates packet collisions within the cluster. Collisions with the members of other clusters are also minimized by the clusterhead's selection of the minimal interference frame.

Ordinary nodes are not static members of clusters, but they choose the cluster they want to join based on the spatial and temporal characteristics of the traffic, taking into account the proximity of the clusterheads and the availability of the data slots within the corresponding cluster. Each frame consists of a control sub-frame for transmission of control packets and a contention-free data sub-frame for data transmission (see Figure 7.2). Beacon packets are used for the announcement of the start of a new frame; Clusterhead Announcement (CA) packets are used for reducing co-frame cluster interference; contention slots are used for initial channel access requests; the header packet is used for announcing the data transmission schedule for the current frame; and Information Summarization (IS) packets are used for announcing the upcoming data packets. IS packets are crucial in energy saving. Each scheduled node transmits its data at the reserved data slot.

In MH-TRACE, nodes switch to sleep mode whenever they are not involved in data transmission or reception, which saves the energy that would be wasted in idle mode or in carrier sensing. Ordinary nodes are in the active mode only during the beacon, header, and IS slots. Furthermore, they stay active for the data slots that they are scheduled to transmit or receive. In addition to these slots, clusterheads stay in the active mode during the CA and contention slots.

Instead of frequency division or code division, MH-TRACE clusters use the same spreading code or frequency, and inter-cluster interference is avoided by using time division among the clusters to enable each node in the network to receive all the desired data packets in its receive range, not just those from nodes in the same cluster. Thus, MH-TRACE clustering does not create hard clusters-the clusters themselves are only used for assigning time slots for nodes to transmit their data.

We modified several features of the original MH-TRACE protocol to integrate it with network-wide broadcasting through flooding. In the original MH-TRACE protocol, which was designed to support single-hop communications, the IS slots are used to measure signal strength, which is used to obtain the approximate distance between the transmitter and the receiver. This is used because packets are discriminated through proximity information, but this technique is not meaningful in network-wide broadcasting. Thus, we modified MH-TRACE by embedding the source ID and the packet sequence number into the IS packet, so that nodes that have already received a particular data packet avoid receiving duplicates of the same packet, which saves a considerable amount of energy.

In the original MH-TRACE protocol there is only one packet drop threshold due to the fact that it supports only single hop communications. However, in network-wide broadcasting many branches of the broadcast tree consist of multiple hops. Applying a single packet drop threshold in each node is not a good strategy, because of the fact that the packets do not need to be dropped until the packet delay exceeds the packet drop threshold. Due to the network dynamics, packet delay is accumulated in time, and a significant portion of the packets are transmitted by the source node at the verge of being dropped. These packets cannot be relayed and are dropped by the neighbors of the source node. The remedy for this problem is to use two packet drop thresholds. At the source node, a smaller packet drop threshold, $T_{drop-source}$, is utilized so that packets that cannot be relayed due to large delays do not waste bandwidth and are automatically dropped by the source node. The rest of the nodes in the network use the standard $T_{drop}$, which is dictated by the application layer. The optimal value of $T_{drop-source}$ is the superframe time, $T_{SF}$. This is because $T_{drop-source}$ should be as low as possible to keep the overall delay as small as possible; and setting $T_{drop-source}$ lower than $T_{SF}$ will cause a packet drop before the next packet arrival, which results in an unutilized data slot.

Characterization of these MAC protocols when they are utilized in network-wide broadcasting through mathematical models is an extremely challenging task. Thus, we opted to investigate their performance through simulations. The simulation environment is presented in the next section.

## 9.3 Simulation Environment

We explored the QoS and energy dissipation characteristics of flooding with the IEEE 802.11, SMAC, and MH-TRACE MAC protocols through extensive ns-2 simulations. We investigated the parameter space with traffic load, node density, and network area/topology as the dimensions. We used a CBR traffic generator with a UDP transport agent to simulate a constant rate voice codec. All the simulations are run for 100 s and repeated three times. We used the propagation and energy models described in Section 2.2.1 and Section 5.1.1,

*Table 9.1.* Constant simulation parameters.

| Acronym | Description | Value |
|---|---|---|
| $C$ | channel rate | 2 Mbps |
| $D_{Tr}$ | transmit range | 250 m |
| $D_{CS}$ | carrier Sense range | 507 m |
| $T_{drop}$ | packet drop threshold | 150 ms |
| $T_{drop-source}$ | packet drop threshold at source | 25 ms |
| $T_{RAD}$ | random assessment delay | 12.5 ms |
| $P_T$ | transmit power | 600 mW |
| $P_R$ | receive power | 300 mW |
| $P_I$ | idle power | 100 mW |
| $P_S$ | sleep power | 10 mW |
| NA | data packet overhead | 10 bytes |
| NA | control packet size | 10 bytes |
| NA | header packet size | 22 bytes |
| $T_{IFS}$ | inter-frame space duration | 16 $\mu s$ |

respectively. Data packet overhead is 10 bytes for IEEE 802.11, SMAC, and MH-TRACE. MH-TRACE control packets are 10 bytes, except the header packet, which is 22 bytes. Acronyms, descriptions and values of the constant parameters used in the simulations are given in Table 9.1.

We used the random way-point mobility model (see Section 2.4) where the node speeds are chosen from a uniform random distribution between 0.0 m/s and 5.0 m/s (the average pace of a marathon runner). In the random way-point mobility model, the average node speed is shown to eventually reach zero for uniform random speed distributions in a [0, vmax] interval [64]. However, throughout the simulation time, average instantaneous node speeds never dropped below 2.3 m/s in any of the scenarios we employed. The pause time is set to zero to avoid non-moving nodes throughout the simulation time. The source node is located in the center of the network. This scenario corresponds to applications where one primary user needs to communicate with all other users in the network; for example, in a battlefield scenario, the commander of a unit (*i.e.*, a squadron) needs to communicate with all the soldiers currently connected to the network.

Although there are many dimensions in ad hoc networks, we limit our study to node density, traffic load, and network area. We examine the traffic load in two regimes: the low traffic regime, which is between 8 Kbps and 32 Kbps, and the high traffic regime, which is 32 Kbps to 128 Kbps. The sampling in the low traffic regime is denser (8 Kbps steps) when compared to the high traffic regime (32 Kbps steps). Traffic (data rate) is changed by varying the packet

*Table 9.2.* Data rate and corresponding data packet payload.

| Regime | Data Rate (Kbps) | Payload (B) | Packet Gen. Period (ms) |
|---|---|---|---|
| | 8 | 50 | 50.0 |
| | 16 | 50 | 25.0 |
| Low | 24 | 75 | 25.0 |
| | 32 | 100 | 25.0 |
| High | 64 | 200 | 25.0 |
| | 96 | 300 | 25.0 |
| | 128 | 400 | 25.0 |

*Table 9.3.* Number of nodes and node density in an 800 m by 800 m network.

| Number of Nodes | Node Density (nodes/km$^2$) |
|---|---|
| 40 | 62.5 |
| 60 | 93.75 |
| 80 | 125 |
| 100 | 156.25 |

size, which is presented in Table 9.2. The main reason for dividing the traffic axis into two parts is that the SMAC protocol can efficiently function only in the low traffic regime. Thus, in the low traffic regime all three of the MAC protocols are evaluated, but in the high traffic regime only IEEE 802.11 and MH-TRACE are evaluated.

Node density is varied between 62.5 nodes per km$^2$ (40 nodes in an 800 m by 800 m area) and 156.25 nodes per km$^2$ (100 nodes in an 800 m by 800 m area) in 31.25 nodes per km$^2$ steps (see Table 9.3). Note that the lowest node density (62.5 nodes/km$^2$) is barely enough to create a connected mobility scenario with the random way-point model. Four different network sizes (and topologies) are utilized in the simulations: 800 m by 800 m, 800 m by 1200 m, 800 m by 1600 m, and 800 m by 2000 m. We use a rectangle shaped network topology (except the 800 m by 800 m network) rather than a square network in order to keep the number of nodes in reasonable limits while increasing the average source/destination path length.

We sampled the traffic-density-area space using eight paths through the parameter space, which we will call sampling paths (see Figure 9.3). The first sampling path represents the variation of data rate (8-32 Kbps) in the low traffic regime while keeping the area (800 m × 800 m) and density (62.5 nodes/km$^2$)

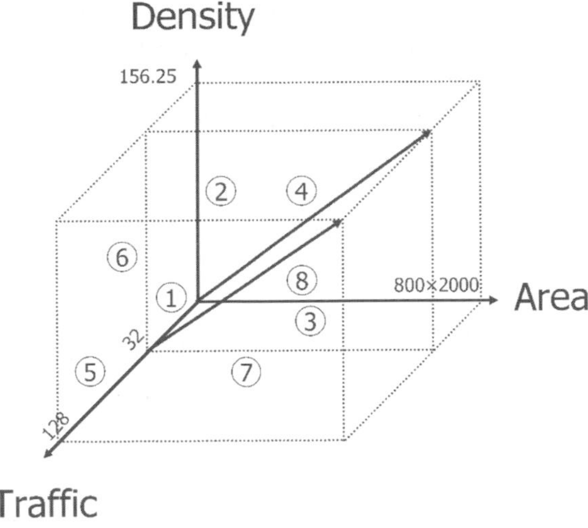


*Figure 9.3.* Sampling the traffic-density-area space.

*Table 9.4.* Data rate, node density, and area for $4^{th}$ and $8^{th}$ paths.

| Point | Data Rate (Kbps) | Node Density (nodes/km$^2$) | Area |
|---|---|---|---|
| S1 | 8 | 62.5 | 800 m × 800 m |
| S2 | 16 | 93.75 | 800 m × 1200 m |
| S3 | 24 | 125 | 800 m × 1600 m |
| S4 | 32 | 156.25 | 800 m × 2000 m |
| S5 | 32 | 62.5 | 800 m × 800 m |
| S6 | 64 | 93.75 | 800 m × 1200 m |
| S7 | 96 | 125 | 800 m × 1600 m |
| S8 | 128 | 156.25 | 800 m × 2000 m |

constant. The second and third sampling paths represent the variation of density (62.5 - 156.25 nodes/km$^2$) and area (800 m × 800 m - 800 m × 2000 m), respectively, while keeping traffic (8 Kbps) and either area (800 m × 800 m) or density (62.5 nodes/km$^2$) constant. The fourth sampling path represents the variation of all parameters, where the network conditions get harsher along the path (see Table 9.4). The fifth, sixth, seventh, and eight sampling paths are the counterparts of the corresponding sampling paths in the high traffic regime.

The metrics that we used for evaluating these protocols are average and minimum packet delivery ratios ($PDR_{Avg}$ and $PDR_{Min}$), packet delay, delay jitter, and energy dissipation. Packet delivery ratio of node $i$ ($PDR_i$) is the ratio of the total number of data packets received by node $i$ to the number of packets generated by the source node. Average PDR is obtained by averaging the PDRs of all the mobile nodes ($N$ mobile nodes in total).

$$PDR_{Avg} = \frac{1}{N} \sum_{i=1}^{N} PDR_i \tag{9.1}$$

Minimum PDR is the PDR of the node with least PDR. Average packet delay at node $i$ ($Delay_{Avg-i}$) is obtained by averaging the delays ($T_j$) of all the packets that are received for the first time at node $i$ ($M_i$), and the global average delay is the average of the delays of $N$ mobile nodes.

$$Delay_{Avg} = \frac{1}{N} \sum_{i=1}^{N} \left( \frac{1}{M_i} \sum_{j} T_j \right) \tag{9.2}$$

RMS delay jitter, which is a measure of the deviation of the packet inter arrival time from the periodicity of the packet generation period, $T_{PG}$, is obtained by using the following equation:

$$Jitter_{RMS} = \sqrt{\frac{1}{N} \sum_{i=1}^{N} \left( \frac{1}{M_i - 1} \sum_{j} (T_j - T_{j-1} - T_{PG})^2 \right)} \tag{9.3}$$

All the energy dissipation results presented in this chapter are the time and ensemble averages, and they are expressed in per node per second energy dissipation form with units mJ/s. Simulation results and analysis are presented in Section 9.4 and Section 9.5.

## 9.4 Low Traffic Regime

### 9.4.1 The First Sampling Path

Data points in the first sampling path are taken along the 8-32 Kbps portion of the traffic axis, where the number of nodes (40 nodes) and network area/topology (800 m × 800 m) is kept constant. IEEE 802.11 performance is summarized in Table 9.5. In the low traffic regime, both the average and the minimum PDR of IEEE 802.11 is almost perfect due to the low level of congestion. The congestion level of the network increases with an increase in the traffic, which is indicated by the increasing number of collisions per transmission with the increasing data rate. However, the number of collisions does not reduce the PDR due to the redundancy of flooding in the low traffic regime.

*Table 9.5.* Simulation results for IEEE 802.11 in the first sampling path (800 m × 800 m network with 40 nodes).

| Data Rate (Kbps) | 8 | 16 | 24 | 32 |
|---|---|---|---|---|
| PDR (avg) | 99% | 99% | 99% | 99% |
| PDR (min) | 99% | 99% | 99% | 99% |
| Delay (ms) | 8 | 8 | 9 | 10 |
| Jitter (ms) | 6 | 5 | 5 | 5 |
| Collision / Trans. | 1.7 | 1.8 | 2.5 | 3.1 |
| Tot Eng / node (mJ/s) | 136.2 | 171.1 | 198.7 | 222.3 |
| Trn Eng / node (mJ/s) | 2.9 (2.1%) | 5.7 (3.3%) | 8.2 (4.1%) | 10.4 (4.7%) |
| Rcv Eng / node (mJ/s) | 19.8 (14.6%) | 39.5 (23.1%) | 55.4 (27.9%) | 69.3 (31.2%) |
| CS Eng / node (mJ/s) | 29.4 (21.6%) | 58.5 (34.2%) | 80.9 (40.7%) | 99.7 (44.9%) |
| Isl Eng / node (mJ/s) | 84.1 (61.7%) | 67.4 (39.4%) | 54.2 (27.3%) | 42.9 (19.3%) |

Even if a packet reception from one rebroadcast node collides, there are many other redundant versions.

Average packet delay is far from the packet drop threshold; however, we see an increasing trend in the packet delay due to the congestion level of the network. Delay jitter, on the other hand, is stable around 5 ms starting with the 16 Kbps data rate. At 8 Kbps data rate, the jitter, 6 ms, is slightly higher than the rest of the data rates, because of the longer inter-arrival time of the data packets at 8 Kbps. There are no dropped packets in IEEE 802.11 in the low traffic regime.

Average energy dissipation per node (Tot E) increases by 63.2% from 8 Kbps to 32 Kbps due to the increase in transmit (Trn E), receive (Rcv E), and carrier sense (CS E) energy dissipation terms in parallel with the increase in the data rate. At 8 Kbps data rate, 83.3% of the total time is spent in the idle mode, which results in 61.7% of the total energy dissipation, whereas at 32 Kbps, 42.5% of the time is spent in the idle mode and only 19.3% of the energy dissipation is spent in the idle mode due to the reduction in the inactive time (*i.e.*, higher data rates result in higher transmit time percentages, which also increase the receive and carrier sense time percentages). The dominant energy dissipation term is carrier sensing at 32 Kbps data rate, which constitutes 44.9% of the total energy dissipation. Although the percentage of transmit energy dissipation is increasing with the data rate, it is still the smallest energy dissipation term. As expected, the ratio of receive and transmit energy dissipations, $6.8 \pm 0.1$, is almost constant for all data rates due to the low level of congestion (*i.e.*, receive/transmit ratio is equal to the average number of neighbors in a collision free network).

*Table 9.6.* Simulation results for SMAC in the first sampling path.

| Data Rate (Kbps) | 8 | 16 | 24 | 32 |
|---|---|---|---|---|
| PDR (avg) | 96% | 96% | 96% | 95% |
| PDR (min) | 95% | 95% | 95% | 95% |
| Delay (ms) | 20 | 16 | 15 | 12 |
| Jitter (ms) | 19 | 13 | 12 | 11 |
| Collision / Trans | 4.0 | 3.9 | 4.1 | 4.2 |
| Drop Pck / s | 3.4 | 4.4 | 6.7 | 15.5 |
| $T_{SMAC}$ (ms) | 50.0 | 25.0 | 25.0 | 25.0 |
| $R_{SMAC}$ | 0.70 | 0.38 | 0.25 | 0.13 |
| Tot E / node (mJ/s) | 72.3 | 136.9 | 176.2 | 206.6 |
| Trn E / node (mJ/s) | 2.8 (3.9%) | 5.6 (4.1%) | 8.0 (4.6%) | 10.0 (4.8%) |
| Rcv E / node (mJ/s) | 18.3 (25.4%) | 36.9 (26.9%) | 51.9 (29.4%) | 63.0 (30.5%) |
| CS E / node (mJ/s) | 26.4 (36.6%) | 53.5 (39.1%) | 73.9 (42.0%) | 88.1 (42.7%) |
| Idl E / node (mJ/s) | 17.6 (24.4%) | 37.1 (27.1%) | 39.8 (22.6%) | 44.2 (21.4%) |
| Slp E/node (mJ/s) | 7.0 (9.8%) | 3.8 (2.8%) | 2.5 (1.4%) | 1.3 (0.6%) |

Simulation results for SMAC in the first data path are shown in Table 9.6. The sleep/active cycle period, $T_{SMAC}$, is matched to the packet generation period, $T_{PG}$, to avoid the excessive interference and contention of sequential data packet waves from the source node. The sleep/active ratio, $R_{SMAC}$, is adjusted to maximize the sleep time while satisfying the QoS requirements of the voice traffic (*i.e.*, minimum PDR is at least 95%), which is the reason that the minimum PDR stays constant. The reason for the monotonic decrease of $R_{SMAC}$ is that the higher $R_{SMAC}$ is not maintainable with an increasing congestion level of the network (induced by the increase in the data rates) without sacrificing QoS. Average packet delay of SMAC is higher than that of IEEE 802.11 due to the sleep periods, where no packet transmissions take place; however, the delay is still much lower than $T_{drop}$. Both delay and jitter decrease with increasing data rate due to shorter sleep periods.

Average energy dissipation of SMAC at 32 Kbps data rate is 186% more than the energy dissipation at 8 Kbps data rate due to the reduction in sleep time, which is utilized in transmit, receive, carrier sense, and idle modes to cope with the higher data rates. SMAC average energy dissipation at 8 Kbps data rate is 47% less than that of IEEE 802.11, which is mainly due to the reduction in the idle energy dissipation (*i.e.*, SMAC idle energy dissipation is 20% of the idle energy dissipation of IEEE 802.11 at 8 Kbps data rate). The major energy dissipation term of SMAC is the carrier sense energy dissipation, and it is unavoidable, because of the fact that carrier sensing is one of the main building blocks of CSMA type medium access control. Receive energy dissipation is

*Table 9.7.* Simulation results for MH-TRACE in the first sampling path.

| Data Rate (Kbps) | 8 | 16 | 24 | 32 |
|---|---|---|---|---|
| PDR (avg) | 99 % | 99 % | 99 % | 99 % |
| PDR (min) | 99 % | 99 % | 99 % | 99 % |
| Delay (ms) | 44 | 45 | 44 | 43 |
| Jitter (ms) | 2 | 2 | 2 | 2 |
| Drop. Pck./s | 75 | 306 | 332 | 523 |
| Data Slt / Sprframe | 70 | 49 | 42 | 35 |
| Tot E / node (mJ/s) | 59.9 | 55.4 | 54.0 | 50.8 |
| Trn E / node (mJ/s) | 4.6 (7.7 %) | 5.9 (10.9 %) | 7.8 (14.3 %) | 8.2 (16.1%) |
| Rcv E / node (mJ/s) | 11.1 (18.7 %) | 10.5 (19.5 %) | 11.9 (21.9 %) | 11.9 (23.4%) |
| CS E / node (mJ/s) | 12.8 (21.5 %) | 11.5 (21.2 %) | 11.1 (20.5%) | 8.2 (16.1%) |
| Idl E / node (mJ/s) | 24.4 (40.8 %) | 18.8 (34.8 %) | 16.0 (29.4 %) | 14.7 (29.0 %) |
| Slp E / node (mJ/s) | 6.8 (11.3 %) | 7.4 (13.7%) | 7.6 (14.0 %) | 7.8 (15.4 %) |

the second largest component of the total energy dissipation, most of which is dissipated on redundant packet receptions. However, in broadcasting it is not possible to discriminate packets due to the lack of RTS/CTS packets (*i.e.*, PAMAS avoids promiscuous listening in unicasting through RTS/CTS packets [21]). Energy savings of SMAC reduces to 7.6% when compared to IEEE 802.11 at 32 Kbps data rate, because of the higher data rate and congestion level of the network. Again, carrier sensing constitutes the largest energy dissipation term and the receive energy dissipation is the second largest energy dissipation term. Transmit energy dissipation never exceeds 5% of the total energy dissipation at any data rate.

MH-TRACE simulation results are presented in Table 9.7. Due to the TDMA structure of MH-TRACE, the length of the data slots should be changed when the data packet length is changed, which results in a change in the number of data slots in each superframe (*i.e.*, superframe length is kept approximately constant, 25.0 ms, thus, larger size data slots result in lower total data slots within a frame and vice versa). For example, there are total of 70 data slots (10 data slots in each of the 7 frames) with 25-byte payload data packets at 8 Kbps data rate and 35 data slots with 100-byte payload data packets at 32 Kbps data rate (see Table 9.8).

MH-TRACE average and minimum PDRs are almost perfect at all data rates (*i.e.*, higher than 99%). However, MH-TRACE packet delay is much higher than both IEEE 802.11 and SMAC due to its superframe structure, where nodes can transmit at most once in one superframe. On the other hand, MH-TRACE jitter is about 60% less than the jitter of IEEE 802.11 for all data rates, which is as important as the average delay in multimedia applications. Reservation

*Table 9.8.* MH-TRACE parameters: Number of frames per superframe, $N_F$, number of data slots per frame, $N_D$, and data packet payload.

| Data Rate | $N_F$ | $N_D$ | Payload |
|---|---|---|---|
| 8 Kbps | 7 | 10 | 25 B |
| 16 Kbps | 7 | 7 | 50 B |
| 24 Kbps | 7 | 6 | 75 B |
| 32 Kbps | 7 | 5 | 100 B |
| 64 Kbps | 7 | 3 | 200 B |
| 96 Kbps | 7 | 2 | 300 B |
| 128 Kbps | 6 | 2 | 400 B |

based channel access is the main reason for such low jitter in MH-TRACE. The average number of dropped packets per second is much higher than the other schemes due to the limited number of data slots (*i.e.*, there is a hard limit on the number of nodes that can have channel access, which is common to all TDMA schemes).

A point worth mentioning is that MH-TRACE is fairly sensitive to the mismatches between the packet generation period, $T_{PG}$, and the superframe time, $T_{SF}$. For example a 1.5% mismatch between $T_{PG}$ and $T_{SF}$ results in 98% and 97% average and minimum PDRs, respectively, and 22 ms packet delay at 32 Kbps data rate. The reason for such behavior is that a certain percentage of the packets, which is approximately equal to the mismatch percentage, are dropped periodically. This also decreases the overall packet delay. Nevertheless, the PDR loss is not high. The packet generation period and the superframe time are matched for the scenarios we present in this chapter.

Analysis of MH-TRACE energy dissipation is a complex task due to its detailed energy conservation mechanisms. In MH-TRACE nodes dissipate energy on both data packets and control packets. For example, when there is no data traffic, the per node energy dissipation of MH-TRACE is 31.6 mJ/s with 8 Kbps configuration, which consists of: (*i*) transmit (11%), receive (6%), and carrier sense (10%) energies dissipated on control packets (*i.e.*, Beacon, Header, *etc.*), (*ii*) idle energy (57%) dissipated during the IS slots (all nodes), and the contention slots (only clusterheads), and (*iii*) sleep mode energy dissipation (26%). When the data traffic is non-zero, nodes dissipate more energy during the IS slots due to the fact that the IS slots are not silent any more (*i.e.*, IS packets are transmitted) and energy dissipation for receive or carrier sensing is three times the energy dissipation for idling.

All of the nodes remain in the active mode during the IS slots, which is the main source of energy dissipation. There is exactly one IS slot for each data slot, and whether the corresponding data slots are utilized or not, all the nodes listen to the IS slots. Actually, this is the mechanism that enables MH-TRACE to avoid receiving redundant packets. For example, if there are 10 data slots in a frame, there are also 10 IS slots in the same frame, and each node should either be receiving all the packets transmitted in the IS slots, waiting in the idle mode, or dissipating energy on carrier sensing. Therefore, the energy dissipation is less if the number of data slots is less. The benefit of dissipating energy in IS slots is that the nodes that monitored the current frame through the IS slots will receive only the data packets that they have not received before. Thus, they will not dissipate energy on redundant data receptions, idle listening, carrier sensing or collisions. Since there are fewer IS slots in the higher data rates than in lower data rates, energy dissipated in the idle and carrier sense modes are lower in higher data rates, which is the reason that the total energy dissipation decreases with increasing data rates.

MH-TRACE energy dissipation at 8 Kbps is 17% less than the energy dissipation of SMAC and 56% less than IEEE 802.11. Despite the fact that SMAC spends slightly more time in the sleep mode at 8 Kbps data rate than MH-TRACE, its total energy dissipation is more than MH-TRACE because of the extra energy dissipation of SMAC in receive and carrier sensing, where MH-TRACE spends most of its active time in the idle mode (idle power is one third of the carrier sense or receive power). At 32 Kbps, MH-TRACE energy dissipation is less than 25% of both SMAC and IEEE 802.11. At 8 Kbps data rate, MH-TRACE transmit energy dissipation is more than 58% higher than both SMAC and IEEE 802.11 due to the extra control packet transmissions. However, at 32 Kbps data rate MH-TRACE transmit energy dissipation is about 80% of the other schemes because of the denied channel access attempts (*i.e.*, the number of data slots are fixed and less than the total number of the nodes in the network).

## 9.4.2 The Second Sampling Path

The number of nodes is increased from 40 to 100 along second sampling path, and the data rate (8 Kbps) and network area/topology (800 m × 800 m) are kept constant. Simulation results for IEEE 802.11 are presented in Table 9.9. Average and minimum PDR (99%), packet delay (8 ms), and delay jitter (6 ms), of IEEE 802.11 is constant for all node densities, which shows that the level of congestion can be handled by IEEE 802.11 in the low traffic regime even with dense networks. However, the increasing trend of the average number of collisions per transmissions hints at the increasing congestion level of the network.

IEEE 802.11 total energy dissipation increases with the increasing node density due to the increase in the receive and carrier sense energy dissipation terms, which is the result of a higher number of nodes in each node's receive

*Table 9.9.* Simulation results for IEEE 802.11 in the second sampling path.

| Number of Nodes | 40 | 60 | 80 | 100 |
|---|---|---|---|---|
| PDR (avg) | 99% | 99% | 99% | 99% |
| PDR (min) | 99% | 99% | 99% | 99% |
| Delay (ms) | 8 | 8 | 8 | 8 |
| Jitter (ms) | 6 | 6 | 6 | 6 |
| Collision /Trans. | 1.7 | 4.2 | 8.3 | 13.4 |
| Tot E / node (mJ/s) | 136.2 | 146.6 | 157.4 | 166.3 |
| Trn E / node (mJ/s) | 2.9 (2.1%) | 2.9 (2.0%) | 2.9 (1.8%) | 2.9 (1.7%) |
| Rcv E / node (mJ/s) | 19.8 (14.6%) | 24.8 (16.9%) | 31.5 (20.0%) | 37.5 (22.6%) |
| CS E / node (mJ/s) | 29.4 (21.6%) | 40.1 (27.3%) | 49.6 (31.5%) | 56.9 (34.2%) |
| Idl E / node (mJ/s) | 84.1 (61.7%) | 78.9 (53.8 %) | 73.5 (46.7%) | 69.1 (41.5%) |

and carrier sense ranges. The transmit energy dissipation does not increase with node density because of the fact that all of the energy entries are normalized with the number of nodes (*i.e.*, per node energy dissipation, per node transmit energy dissipation, *etc.*).

Simulation results for SMAC in the second sampling path are presented in Table 9.10. We kept the minimum PDR of SMAC fixed by varying RSMAC, which resulted in shortened sleep periods at higher node densities. Both delay and jitter decrease with the increasing node density due to the shortened sleep period. Nevertheless, the congestion level of the network increases with node density, which manifests itself with the increasing trend in packet drops per second and the average number of data packet collisions per transmission.

Average energy dissipation of SMAC is 52% of the energy dissipation of IEEE 802.11 in the 40-node network. This ratio increases to 81% for the 100-node network. The reduction in energy savings is due to the increase in receive, carrier sense, and idle energy dissipation.

MH-TRACE simulation results in the second sampling path are presented in Table 9.11. The average and minimum PDR, packet delay and delay jitter of MH-TRACE are almost constant for all node densities. Like in the first sampling path, the average packet delay of MH-TRACE is higher than both IEEE 802.11 and SMAC in the second sampling path. The number of dropped packets per second increases with increasing node density due to the fact that the higher number of nodes cannot all gain channel access in denser networks. However, note that this does not affect the PDR, due to the redundancy inherent in the flooding protocol.

Average per node energy dissipation of MH-TRACE is $62.3 \pm 2.3$ mJ/s for all node densities. Per node transmit energy decreases with node density because the ratio of the data transmissions per node decreases with the node

*Table 9.10.* Simulation results for SMAC in the second sampling path.

| Number of Nodes | 40 | 60 | 80 | 100 |
|---|---|---|---|---|
| PDR (avg) | 96% | 96% | 96% | 96% |
| PDR (min) | 95% | 95% | 95% | 95% |
| Delay (ms) | 20 | 18 | 16 | 13 |
| Jitter (ms) | 19 | 11 | 11 | 9 |
| Coll / Trans. | 4.0 | 5.6 | 10.1 | 14.0 |
| Drop Pck / s | 3.4 | 3.8 | 4.5 | 5.3 |
| RSMAC | 0.70 | 0.60 | 0.50 | 0.40 |
| Tot E / node (mJ/s) | 72.3 | 94.2 | 115.2 | 134.8 |
| Trn E / node (mJ/s) | 2.8 (3.9%) | 2.8 (2.9%) | 2.8 (2.5%) | 2.8 (2.1%) |
| Rcv E / node (mJ/s) | 18.3 (25.4%) | 23.7 (25.9%) | 30.6 (26.1%) | 36.9 (27.3%) |
| CS E / node (mJ/s) | 26.4 (36.6%) | 37.5 (39.6%) | 47.1 (41%) | 55.5 (41.2%) |
| Idl E / node (mJ/s) | 17.6 (24.4%) | 24.2 (25.6%) | 29.6 (26%) | 35.6 (26.4%) |
| Slp E / node (mJ/s) | 7.0 (9.8%) | 6.0 (6.4%) | 5.0 (4.4%) | 4.0 (3.0%) |

*Table 9.11.* Simulation results for MH-TRACE in the second sampling path.

| Number of Nodes | 40 | 60 | 80 | 100 |
|---|---|---|---|---|
| PDR (avg) | 99% | 99% | 99% | 99% |
| PDR (min) | 99% | 99% | 99% | 99% |
| Delay (ms) | 44 | 46 | 46 | 45 |
| Jitter (ms) | 2 | 2 | 2 | 2 |
| Drop Pck / s | 75 | 219 | 663 | 1292 |
| Tot E / node (mJ/s) | 59.9 | 62.9 | 64.1 | 62.6 |
| Trn E / node (mJ/s) | 4.6 (7.7%) | 4.4 (7.0%) | 3.8 (5.9%) | 3.2 (5.2%) |
| Rcv E / node (mJ/s) | 11.1 (18.7%) | 12.7 (20.2%) | 14.5 (22.5%) | 14.8 (23.7%) |
| CS E / node (mJ/s) | 12.8 (21.5%) | 16.8 (26.8%) | 19.0 (29.6%) | 19.5 (31.2%) |
| Idl E / node (mJ/s) | 24.4 (40.8%) | 22.1 (35.2%) | 20.0 (31.2%) | 18.0 (28.9%) |
| Slp E / node (mJ/s) | 6.8 (11.3%) | 6.8 (10.8%) | 6.9 (10.8%) | 7.0 (11.1%) |

density (*i.e.*, the number of data transmissions do not increase as fast as the node density). Actually, the number of data slots does not change significantly when the network area is kept constant because the number of clusterheads is primarily determined by the network area, and the total number of data slots per clusterhead is constant. However, in low density networks, utilization of the data slots of the outer clusterheads is not as high as the utilization of the inner clusterheads. Thus, the number of data slots in use is higher for denser networks, although the number of data slots is not necessarily higher.

Both the actual and the percentage contribution of receive and carrier sense energy dissipations increase, and the contribution of the idle and transmit energy dissipations decrease, due to the decrease in the number of transmissions per node with increasing node density (*i.e.*, utilization of the data slots, especially the data slots in the outer parts of the network, increase with the node density). There is a slow increase in the sleep energy dissipation due to the reduction of the ratio of the clusterheads to total number of nodes, which have more time to sleep (*i.e.*, ordinary nodes do not need to stay in the active mode during the contention slots).

### 9.4.3 The Third Sampling Path

Along the third sampling path, network area/topology is varied from 800m × 800 m to 800 × 2000 m while keeping the data rate (8 Kbps) and node density (62.5 nodes/km$^2$) constant. The purpose of this sampling path is to reveal the effects of path length on network performance. Simulation results for IEEE 802.11, SMAC, and MH-TRACE are summarized in Table 9.12, Table 9.13, and Table 9.14, respectively. Since energy consumption is not significantly affected from the variations in the path length, we do not include the detailed energy dissipation results in these tables.

IEEE 802.11 PDR is not affected by the variations in path length in the low traffic regime, and it is stable (around 99%) for the path lengths we investigated in the third sampling path. Packet delay and delay jitter increase linearly with the path length from 8 ms and 6 ms to 16 ms and 7 ms, respectively. IEEE 802.11 energy dissipation per node does not change significantly and stabilizes around 140 mJ/s.

After the initial reduction from 0.70 to 0.60, SMAC sleep/active ratio stays constant at 0.60. Average packet delay of SMAC increases from 20 ms to 50 ms with increasing average path length while the delay jitter varies from 19 ms to 30 ms. Energy dissipation of SMAC is in parallel with the sleep/active ratio. The behavior of IEEE 802.11 and SMAC do not change significantly, except the packet delay and jitter, due to the fact that the delay in these medium access schemes is not high enough to affect PDR with the low level of congestion.

Average PDR of MH-TRACE is above 95% for all network topologies; however, minimum PDR drops below 95 % starting with the 800 m × 1600 m network. The reason for such low PDR is the high packet delay of MH-TRACE, which is indicated by the average packet delay in Table 9.14. The nodes with low PDRs are located far from the source node, which is located at the center of the network. On the other hand, MH-TRACE delay jitter is still less than half of the delay jitter obtained with IEEE 802.11 and is about 10% of SMAC delay jitter. MH-TRACE energy dissipation per node stays in a narrow band around 60 mJ/s, which is 63% less than the IEEE 802.11 energy dissipation and 32% less than the SMAC energy dissipation for the 800 m × 2000 m network.

*Table 9.12.* Simulation results for IEEE 802.11 in the third sampling path.

| Area (Topology) | 800 × 800 | 800 × 1200 | 800 × 1600 | 800 × 2000 |
|---|---|---|---|---|
| PDR (avg) | 99% | 99% | 99% | 99% |
| PDR (min) | 99% | 99% | 99% | 99% |
| Delay (ms) | 8 | 10 | 12 | 16 |
| Jitter (ms) | 6 | 6 | 7 | 7 |
| Tot E / node (mJ/s) | 136.2 | 141.2 | 140.4 | 140.5 |

*Table 9.13.* Simulation results for SMAC in the third sampling path.

| Area (Topology) | 800 × 800 | 800 × 1200 | 800 × 1600 | 800 × 2000 |
|---|---|---|---|---|
| PDR (avg) | 96% | 96% | 96% | 96% |
| PDR (min) | 95% | 95% | 95% | 95% |
| Delay (ms) | 20 | 27 | 41 | 50 |
| Jitter (ms) | 19 | 20 | 25 | 30 |
| $R_{SMAC}$ | 0.70 | 0.60 | 0.60 | 0.60 |
| Tot E / node (mJ/s) | 72.3 | 87.6 | 87.8 | 88.2 |

*Table 9.14.* Simulation results for MH-TRACE in the third sampling path.

| Area (Topology) | 800 × 800 | 800 × 1200 | 800 × 1600 | 800 × 2000 |
|---|---|---|---|---|
| PDR (avg) | 99% | 99% | 99% | 97% |
| PDR (min) | 99% | 99% | 92% | 67% |
| Delay (ms) | 44 | 54 | 73 | 89 |
| Jitter (ms) | 2 | 2 | 3 | 3 |
| Tot E / node (mJ/s) | 59.9 | 62.0 | 60.8 | 60.5 |

### 9.4.4 The Fourth Sampling Path

Data points in the fourth sampling path are taken along the diagonal of the low traffic regime parameter space, where S$i$ stands for the samples on the path (*i.e.*, the first row of Table 9.4 is S1, the second row of Table 9.4 is S2, and so on). Simulation results obtained along the fourth sampling path for IEEE 802.11, SMAC, and MH-TRACE are presented in Table 9.15, Table 9.16, and Table 9.17, respectively.

Average and minimum PDRs of IEEE 802.11 drop below 95% starting with S3, because of the high congestion level of the network. Packet delay and

*Table 9.15.* Simulation results for IEEE 802.11 in the fourth sampling path.

| | S1 | S2 | S3 | S4 |
|---|---|---|---|---|
| PDR (avg) | 99% | 99% | 92% | 80% |
| PDR (min) | 99% | 99% | 91% | 76% |
| Delay (ms) | 8 | 11 | 33 | 58 |
| Jitter (ms) | 6 | 6 | 12 | 15 |
| Tot E / node(mJ/s) | 136.2 | 192.9 | 251.3 | 292.0 |

*Table 9.16.* Simulation results for SMAC in the fourth sampling path.

| | S1 | S2 | S3 | S4 |
|---|---|---|---|---|
| PDR (avg) | 96% | 96% | | |
| PDR (min) | 95% | 95% | | |
| Delay (ms) | 20 | 17 | | |
| Jitter (ms) | 19 | 11 | | |
| RSMAC | 0.70 | 0.28 | | |
| Tot E / node (mJ/s) | 72.3 | 173.4 | | |

*Table 9.17.* Simulation results for MH-TRACE in the fourth sampling path.

| | S1 | S2 | S3 | S4 |
|---|---|---|---|---|
| PDR (avg) | 99% | 99% | 99% | 98% |
| PDR (min) | 99% | 99% | 97% | 96% |
| Delay (ms) | 44 | 65 | 77 | 88 |
| Jitter (ms) | 2 | 2 | 3 | 3 |
| Tot E / node (mJ/s) | 59.9 | 56.9 | 54.3 | 53.1 |

jitter also increase along the sampling path. Node density and data rate are the dominant factors affecting the congestion level of the network. Although IEEE 802.11 does not exhibit a significant QoS deterioration in low density and high data rate networks (*i.e.*, S1) or high density and low data rate networks (*i.e.*,S2), when we combine high node density and high data rate, the resultant congestion level of the network is more than that can be handled by the contention resolution mechanism of IEEE 802.11. Energy dissipation of IEEE 802.11 increases along the sampling path due to the increase in the total number (node density) and size (data rate) of data packet transmissions.

Since the performance of IEEE 802.11 is below the QoS requirements for the second half of the fourth sampling path, it is not meaningful to try to save energy, which would further deteriorate the QoS. Thus, SMAC results are presented only for the first half of the fourth sampling path. The sleep/active ratio of SMAC drops from 0.70 at S1 to 0.28 at S2. SMAC energy dissipation is 10% less than that of IEEE 802.11 at S2.

Surprisingly, MH-TRACE minimum PDR is above 95% all along the fourth sampling path, unlike the third sampling path, where MH-TRACE minimum PDR drops below 95% in the second half of the third sampling path. By investigating the paths traversed by the packets, we found that the average number of hops from the source to the mobile nodes decreases with node density due to the increase in connectivity (*i.e.*, average degree of a node). MH-TRACE packet delay at highly congested networks is comparable with the packet delay of IEEE 802.11 (*i.e.*, MH-TRACE packet delay is 50% more than IEEE 802.11 packet delay at S4), while IEEE 802.11 packet delay is significantly lower than MH-TRACE delay at lightly loaded networks (*i.e.*, IEEE 802.11 packet delay at S1 is less than 20% of MH-TRACE packet delay).

The decrease in the per node energy dissipation of MH-TRACE is mainly due to the increase in node density and decrease in the number of data slots, which are explained in detail in Section 9.4.1. MH-TRACE energy dissipation is less than a third of the energy dissipation of SMAC at S2, and at S4 MH-TRACE energy dissipation is less than a fifth of the energy dissipation of IEEE 802.11.

## 9.5 High Traffic Regime

Having completed the analysis of the sampling paths within the low traffic regime, starting with the fifth sampling path we focus on the high traffic regime. In the high traffic regime we investigate IEEE 802.11 and MH-TRACE only, because beyond the 32 Kbps data rate it is not possible to save any energy with SMAC, which is its main feature.

### 9.5.1 The Fifth Sampling Path

Data points in the fifth sampling path are taken along the 32-128 Kbps portion of the traffic axis with the number of nodes (40 nodes) and network area/topology (800 m × 800 m) kept constant. Unlike the first sampling path, where IEEE 802.11 PDR stays constant at 99%, both the average and minimum PDR of IEEE 802.11 drops with increasing data rate (see Table 9.18) due to severe congestion. Note that despite the fact that the PDR decreases with increasing data rate, throughput (*i.e.*, number of bytes) increases with the increasing data rate. For example, the amount of data relayed to the minimum PDR node at 32 Kbps node is 4 Kbytes per second (*i.e.*, 32 Kbps), whereas at 128 Kbps data rate the amount of data conveyed to the minimum PDR node is

*Table 9.18.* Simulation results for IEEE 802.11 in the fifth path.

| Data Rate (Kbps) | 32 | 64 | 96 | 128 |
|---|---|---|---|---|
| PDR (avg) | 99% | 89 % | 82% | 78% |
| PDR (min) | 99% | 89 % | 64% | 39% |
| Delay (ms) | 10 | 31 | 54 | 68 |
| Jitter (ms) | 5 | 15 | 20 | 22 |
| Coll / Trans | 3.1 | 7.3 | 6.7 | 5.5 |
| Drop Pck / s | 0.0 | 9.3 | 250.2 | 388.3 |
| Tot E / node (mJ/s) | 222.3 | 273.6 | 284.5 | 289.7 |

*Table 9.19.* Simulation results for MH-TRACE in the fifth path.

| Data Rate (Kbps) | 32 | 64 | 96 | 128 |
|---|---|---|---|---|
| PDR (avg) | 99% | 99% | 99% | 99% |
| PDR (min) | 99% | 99% | 93% | 89% |
| Delay (ms) | 43 | 44 | 45 | 44 |
| Jitter (ms) | 2 | 2 | 2 | 2 |
| Drop Pck / s | 523 | 907 | 1234 | 1199 |
| Tot E / node (mJ/s) | 50.8 | 48.8 | 45.4 | 44.4 |

*Table 9.20.* Simulation results for IEEE 802.11 in the fifth path with $T_{drop} \rightarrow \infty$.

| Data Rate (Kbps) | 32 | 64 | 96 | 128 |
|---|---|---|---|---|
| PDR (avg) | 99% | 88% | 74% | 59% |
| PDR (min) | 99% | 88% | 73% | 59% |
| Delay (ms) | 10 | 31 | 1798 | 3152 |
| Jitter (ms) | 5 | 15 | 23 | 29 |

*Table 9.21.* Simulation results for MH-TRACE in the fifth path with $T_{drop} \rightarrow \infty$.

| Data Rate (Kbps) | 32 | 64 | 96 | 128 |
|---|---|---|---|---|
| PDR (avg) | 99% | 99% | 99% | 99% |
| PDR (min) | 99% | 99% | 99 % | 89% |
| Delay (ms) | 191 | 226 | 285 | 312 |
| Jitter (ms) | 2 | 2 | 3 | 3 |

6.25 Kbytes per second. The decrease in the average number of collisions per transmission is due to the decrease in the number of data packet transmissions. Despite the fact that the number of data transmissions decreases with increasing data rate, transmit, receive, and carrier sense energies increase due to the increase in the size of the data packets.

MH-TRACE average PDR stays constant at 99% in the fifth data path (see Table 9.19). However, minimum PDR drops below 95% for data rates higher than 64 Kbps. The relatively low number of data slots per superframe is the reason for the low minimum PDR at higher data rates. Since the network layer algorithm is flooding, there is no coordination in relaying the data packets (*i.e.*, statistical multiplexing). When the number of rebroadcasts is low (limited number of data slots per superframe) failure of the formation of a dominating set in some broadcast waves is inevitable. Since the average number of clusterheads is constant for all data rates (*i.e.*, average number of clusterheads is mainly determined by the network size), the number of data slots in the network is determined by the number of data slots per frame (*i.e.*, total number of available data slots in the network is the product of the number of clusterheads and the number of data slots per frame). Thus, some of the nodes, especially the ones far from the source node, have relatively low PDR compared with the rest of the network.

MH-TRACE packet delay and jitter do not change significantly along the fifth sampling path. MH-TRACE energy dissipation exhibits a slight decrease along the fifth sampling path due to the decrease in the number of IS slots, as was described in Section 9.4.2. MH-TRACE energy dissipation is less than one sixth of the energy dissipation of IEEE 802.11 at 128 Kbps data rate.

Actually, the PDR of IEEE 802.11 is higher in highly congested networks (>64 Kbps) if there is a hard constraint on the maximum packet delay (*i.e.*, packets with delays higher than $T_{drop}$). Table 9.20 and Table 9.21 present the simulation results for IEEE 802.11 and MH-TRACE along the fifth sampling path with no packet drop threshold (*i.e.*, $T_{drop} \rightarrow \infty$), respectively. At 96 Kbps and 128 Kbps data rates, average PDR of IEEE 802.11 with packet drops is larger than the case with no packet drops, yet the minimum PDR is higher without packet drops. This is because the average PDR is primarily affected by the congestion level of the network and the difference between the average and minimum PDRs is due to the delay constraint. MH-TRACE PDR is not affected significantly by the packet drop threshold. However, the packet delay rises to formidably high levels, yet still is a magnitude lower than the IEEE 802.11 packet delay in high congestion (data rate > 64 Kbps).

## 9.5.2 The Sixth Sampling Path

The number of nodes is increased along the sixth sampling path, while keeping the data rate (32 Kbps) and network area (800 m × 800 m) constant.

*Table 9.22.* Simulation results for IEEE 802.11 in the sixth path.

| Number of Nodes | 40 | 60 | 80 | 100 | |
|---|---|---|---|---|---|
| PDR (avg) | 99% | 94% | 88% | 77% | |
| PDR (min) | 99%Delay (ms) | 10 | 17 | 28 | 33 |
| Jitter (ms) | 5 | 7 | 13 | 15 | |
| Tot E / node | (mJ/s) | 222.3 | 240.4 | 246.5 | 247.8 |

*Table 9.23.* Simulation results for MH-TRACE in the sixth path.

| Number of Nodes | 40 | 60 | 80 | 100 |
|---|---|---|---|---|
| PDR (avg) | 99% | 99% | 99% | 99% |
| PDR (min) | 99% | 99% | 99% | 99% |
| Delay (ms) | 43 | 41 | 44 | 42 |
| Jitter (ms) | 2 | 2 | 2 | 2 |
| Tot E / node (mJ/s) | 50.8 | 51.4 | 50.7 | 49.9 |

Table 9.22 and Table 9.23 present the simulation results for IEEE 802.11 and MH-TRACE, respectively. IEEE 802.11 average PDR drops below 95% starting with the 60 node network, and reaches 77% for the 100 node network. Decrease of the PDR and increase of the packet delay and delay jitter are all due to the increase in the congestion level of the network with increasing node density. There is not a significant gap between the average and minimum PDRs of IEEE 802.11 due to the comparatively lower packet delays when compared to the packet delays along the fifth sampling path.

Both the average and minimum PDR of MH-TRACE stay constant at 99%, and the packet delay also lies in a narrow band around 43 ms. MH-TRACE energy dissipation at 156.25 nodes/km$^2$ node density is approximately one fifth of the energy dissipation of IEEE 802.11.

### 9.5.3 The Seventh Sampling Path

Data points along the seventh sampling path are taken by varying the network size from 800 m × 800 m to 800 m × 2000 m, while keeping the data rate (32 Kbps) and node density (62.5 nodes/km$^2$) constant. IEEE 802.11 PDR stays above 99% all along the seventh sampling path (Table 9.24). However, the increase in average packet delay shows that the PDR will start to decrease for longer path lengths. MH-TRACE minimum PDR also drops below 95% in the second half of the sampling path due to the packet drops arising because of

*Table 9.24.* Simulation results for IEEE 802.11 in the seventh path.

| Area (Topology) | 800 × 800 | 800 × 1200 | 800 × 1600 | 800 × 2000 |
|---|---|---|---|---|
| PDR (avg) | 99% | 99% | 99% | 99% |
| PDR (min) | 99% | 99% | 99% | 98% |
| Delay (ms) | 10 | 19 | 33 | 58 |
| Jitter (ms) | 5 | 8 | 11 | 15 |
| Tot E / node (mJ/s) | 222.3 | 235.4 | 251.3 | 252.8 |

*Table 9.25.* Simulation results for MH-TRACE in the seventh path.

| Area (Topology) | 800 × 800 | 800 × 1200 | 800 × 1600 | 800 × 2000 |
|---|---|---|---|---|
| PDR (avg) | 99% | 99% | 88% | 88% |
| PDR (min) | 99% | 99% | 40% | 26% |
| Delay (ms) | 43 | 52 | 71 | 86 |
| Jitter (ms) | 2 | 2 | 3 | 4 |
| Tot E / node (mJ/s) | 50.8 | 52.5 | 52.9 | 53.4 |

the longer paths between the source and the distant nodes (Table 9.25). MH-TRACE average and minimum PDRs in the seventh sampling path are lower than their counterparts in the third sampling path because of the fact that the total number of data slots in the higher data rate networks is lower than total number of data slots in the lower data rate networks, which deteriorates the path diversity and consequently increases the packet delay.

### 9.5.4 The Eighth Sampling Path

Data points in the eighth sampling path are taken along the diagonal of the high traffic regime parameter space, where Si stand for the samples on the path (see Table 9.4). Simulation results obtained along the eighth sampling path for IEEE 802.11 and MH-TRACE are presented in Table 9.26 and Table 9.27. In the eighth sampling path, which is the most challenging in this study, both IEEE 802.11 and MH-TRACE failed to maintain a minimum PDR of 95% after the first sample on the path. Congestion is the main reason for such deterioration of IEEE 802.11 due to the increase in the data rate and node density, which means a higher number of larger data packets. The main reason for the deterioration of MH-TRACE performance is the high packet delays due to the increase in average path length and the reduction of the total number of data slots per km$^2$ along the eighth sampling path. Although the average PDR of MH-TRACE is

*Table 9.26.* Simulation results for IEEE 802.11 in the eighth sampling path.

| | S5 | S6 | S7 | S8 |
|---|---|---|---|---|
| PDR (avg) | 99% | 88% | 74% | 64% |
| PDR (min) | 99% | 76% | 35% | 33% |
| Delay (ms) | 10 | 90 | 98 | 116 |
| Jitter (ms) | 6 | 24 | 23 | 24 |
| Tot E / node (mJ/s) | 222.3 | 272.3 | 281.2 | 267.5 |

*Table 9.27.* Simulation results for MH-TRACE in the eighth sampling path.

| | S5 | S6 | S7 | S8 |
|---|---|---|---|---|
| PDR (avg) | 99% | 98% | 90% | 84% |
| PDR (min) | 99% | 90% | 40% | 15% |
| Delay (ms) | 43 | 71 | 90 | 106 |
| Jitter (ms) | 2 | 2 | 2 | 2 |
| Tot E / node (mJ/s) | 50.8 | 49.6 | 41.8 | 46.2 |

higher than IEEE 802.11 along the eighth sampling path, the minimum PDR of MH-TRACE is lower than that of IEEE 802.11 at the fourth sampling point due to the excessive packet drops at locations close to the edges of the network. Furthermore, IEEE 802.11 delay is higher than that of MH-TRACE at the fourth sampling point due to the high level of congestion. We present a summary of all of these simulations and analysis in the following section.

## 9.6 Summary

In this chapter we investigated the role of medium access control on the QoS and energy dissipation characteristics of network-wide real-time data broadcasting through flooding using three MAC protocols (IEEE 802.11, SMAC, and MH-TRACE) within the data rate, node density, and network area/topology parameter space. The ranges of the parameter space are chosen to characterize the behavior of the broadcast architectures. Thus, we identified the breaking points of each MAC layer in flooding.

IEEE 802.11 achieves almost perfect PDR in low density networks (where the number of nodes is barely enough to create a connected network with the RWP mobility model with pedestrian speed) with low (8 Kbps) to medium (32 Kbps) data rates. However, for higher data rates (*i.e.*, data rates higher than 32 Kbps), IEEE 802.11 PDR exhibits a sharp decrease due to the high level of congestion. In low data traffic networks (8 Kbps), IEEE 802.11 is capable of handling low (62.5 nodes/km$^2$) to high (156.25 nodes/km$^2$) node densities without sacrificing the PDR. For high data rates (>32 Kbps ), even with low node density IEEE 802.11 cannot maintain the network stability and PDR deteriorates significantly. IEEE 802.11 is virtually immune to changes in the average path length (*i.e.*, for the path lengths we considered in this study) for low node densities and low data rates because of its relatively lower packet delay. However, there is a limit on the serviceable maximum path length, which is determined by the delay limit of the application (*i.e.*, $T_{drop}$). IEEE 802.11 performance is affected seriously by the combined high node density and high data rates, which also limits the path length scalability of IEEE 802.11. Energy dissipation of IEEE 802.11 is determined mainly by the total number of packets transmitted, and there is no built-in energy saving mechanisms for IEEE 802.11 in the ad hoc mode of operation.

The main advantage of SMAC is its capability of saving energy wasted in the idle mode by the underlying IEEE 802.11 protocol. SMAC successfully saves energy in low node density and low data traffic networks without sacrificing the QoS requirements of the application. However, with increasing node densities and/or data rates, SMAC energy savings diminishes quickly. For medium node density and low data rate networks, SMAC energy savings are only marginal due to the limited sleep time. The same applies to low node density and medium data rate networks for SMAC. Although SMAC packet delay and delay jitter is higher than IEEE 802.11, it can successfully meet the QoS requirements of the application for longer path lengths in low node density and low data rate networks. SMAC cannot operate effectively in the high data regime (>32 Kbps), because the underlying IEEE 802.11 needs all the bandwidth available to avoid congestion; thus, there is no bandwidth available to waste in the sleep mode to save energy.

MH-TRACE can maintain 99% PDR up to medium-high (64 Kbps) data rates in low density networks. Under all node densities with low (8 Kbps) and medium (32 Kbps) data rates, MH-TRACE is capable of maintaining the QoS requirements of the application due to its coordinated channel access mechanism. However, due to its high packet delay, MH-TRACE cannot maintain the required minimum PDR in large networks. However, in combined difficulty levels (low-medium node densities and data rates) MH-TRACE QoS metrics are better than the other schemes. MH-TRACE energy dissipation is significantly lower than the other schemes for the entire parameter space due to its schedule based channel access and data discrimination mechanisms.

Chapter 10

# NB-TRACE PROTOCOL ARCHITECTURE

## 10.1 Introduction

In Chapter 9 we presented a comparative analysis of MH-TRACE-based flooding and other flooding architectures. Although MH-TRACE-based flooding energy dissipation is much lower than the other schemes, all of the flooding architectures (including MH-TRACE-based flooding) have low spatial reuse efficiency due to the redundancy of flooding as a network layer broadcast technique. Thus, the need for a network layer broadcast architecture, which inherits the energy efficiency of MH-TRACE and combines it with spatial reuse efficiency, is obvious.

All of the major components of energy and spatial reuse efficient QoS-supporting network-wide broadcasting have been investigated in the literature [109, 10, 20, 22, 93]. However, a multi-objective architecture that integrates all of the design goals has not been proposed. In this chapter, we present such an architecture, called Network-wide Broadcasting through Time Reservation using Adaptive Control for Energy efficiency (NB-TRACE) [49, 51].

The remainder of this chapter is organized as follows. Section 10.2 describes the NB-TRACE architecture. The simulation environment and results are presented in Section 10.3. A summary of this chapter is presented in Section 10.4.

## 10.2 Protocol Architecture

### 10.2.1 Design Principles

Since we wanted to keep the MH-TRACE structure intact, we followed a bottom up approach to design the network layer architecture, rather than a top down approach (*i.e.*, the network layer is tailored according to the MAC layer). We considered combining MH-TRACE and an existing network layer broadcast algorithm to achieve energy-efficient network-wide broadcasting of voice data.

Due to its simplicity, we first integrated flooding with MH-TRACE. In MH-TRACE-based flooding each node that can obtain channel access continuously rebroadcasts the voice packets. In network-wide broadcasting we employ the IS slots of MH-TRACE to transmit the unique ID of the corresponding voice packets (*i.e.*, the source node ID and data packet sequence number constitutes a unique ID). Thus nodes in the receive range of the transmitting node are informed ahead of time about upcoming data transmissions and avoid receiving multiple copies of the same packet, which saves a considerable amount of energy. However, due to the inherent inefficiencies of flooding, spatial reuse of the combined architecture was not satisfactory (*i.e.*, too many redundant rebroadcasts). On the other hand, the energy dissipation of MH-TRACE-based flooding was far better than flooding with other MAC protocols (see Chapter 9).

The other network layer broadcasting algorithms were not easy to integrate with MH-TRACE without degrading system performance due to the application-specific design of MH-TRACE. For example, when we use gossiping in conjunction with MH-TRACE, due to the per packet based probabilistic channel access, the reservation mechanism of MH-TRACE cannot function properly (*i.e.*, continuous utilization of the data slots is necessary). Furthermore, the advantageous features of MH-TRACE (*e.g.*, organization of the network into clusters, automatic renewal of channel access) cannot be fully utilized by any existing network-layer broadcasting algorithm. Thus, there is a need for a new application-specific network layer algorithm integrated with an application-specific MAC layer (*i.e.*, NB-TRACE).

The main function of NB-TRACE is to connect the Non-Connected Dominating Set (NCDS) formed by the clusterheads (CHs), maintained by the underlying MH-TRACE protocol. This mostly eliminates the burden of maintaining a CDS by the network layer because the maintenance of the cluster structure is done by the MAC layer, which clearly is a benefit of cross-layer design.

The basic design philosophy of NB-TRACE is to flood the network and, by using the properties of the underlying MH-TRACE architecture, to prune the network as much as possible while maintaining a connected dominating set with minimal control packet exchange (*i.e.*, minimizing the overhead). We also wanted to keep the data slots exclusively for data packets rather than using them for control packets in order to not interrupt data streams.

### 10.2.2 Overview

In NB-TRACE, the network is organized into overlapping clusters, each managed by a CH. Channel access is granted by the CHs through a dynamic, distributed Time Division Multiple Access (TDMA) scheme, which is organized into periodic superframes. Initial channel access is though contention; however, a node that utilizes the granted channel access automatically reserves a data slot in the subsequent superframes. The superframe length, $T_{SF}$, is matched to the periodic rate of voice generation, $T_{PG}$.

Data packets are broadcast to the entire network through flooding at the beginning of each data session. Each rebroadcasting (relay) node explicitly acknowledges (ACKs) the upstream node as part of its data transmission. Relay nodes that do not receive any ACK in $T_{ACK}$ time cease to rebroadcast. As an exception, the CHs continue to rebroadcast regardless of any ACK, which prevents the eventual collapse of the broadcast tree. Due to node mobility, the initial tree will be broken in time, so NB-TRACE is equipped with several mechanisms to maintain the broadcast tree over time.

In NB-TRACE, a broadcast tree is formed by the initiation of a source node, however, once a tree is formed (*i.e.*, once nodes determine their roles), then other sources do not need to create another tree; instead, they use the existing organization to broadcast their packets.

NB-TRACE broadcasting and packet flow is illustrated in Figure 10.1. NB-TRACE is composed of five basic building blocks: (*i*) Initial Flooding (IFL), (*ii*) Pruning (PRN), (*iii*) Repair Branch (RPB), (*iv*) Create Branch (CRB), and (*v*) Activate Branch (ACB). The NB-TRACE algorithm flowchart is presented in Figure 10.2. Actually, all of theses building blocks are functioning simultaneously; however, we describe them as sequential mechanisms to make them easier to understand.

### 10.2.3 Initial Flooding (IFL)

A source node initiates a session by broadcasting packets to its one-hop neighbors. Nodes that receive a data packet contend for channel access, and the ones that obtain channel access retransmit the data they received. Eventually, the data packets are received by all the nodes in the network, possibly multiple times. Each rebroadcasting node ACKs its upstream node by announcing the ID of its upstream node in its IS packet, which precedes its data packet transmission (see Figure 7.2). A source node announces its own ID as its upstream node. Source node ID, Flow ID, Packet ID, CH Status, and IFL ID are also announced in the IS packet (see Figure 10.3). At the beginning of a broadcast session IFL ID is set to one for $T_{IFL}$ time to force the nodes to switch to active mode. At this point, some of the nodes have multiple upstream nodes. A node with multiple upstream nodes chooses the upstream node that has the least packet delay as its upstream node to be announced in its IS packet in order to minimize the delay.

Whenever a source node stops its broadcast, it sets the Packet ID field to null-ID. Nodes that receive an IS packet with a null Packet ID mark the corresponding source and flow IDs as inactive for future reference. Furthermore, all the nodes record the time instants that they last received an IS packet with the Upstream Node ID set as their own ID from any other node (*i.e.*, ACK reception).

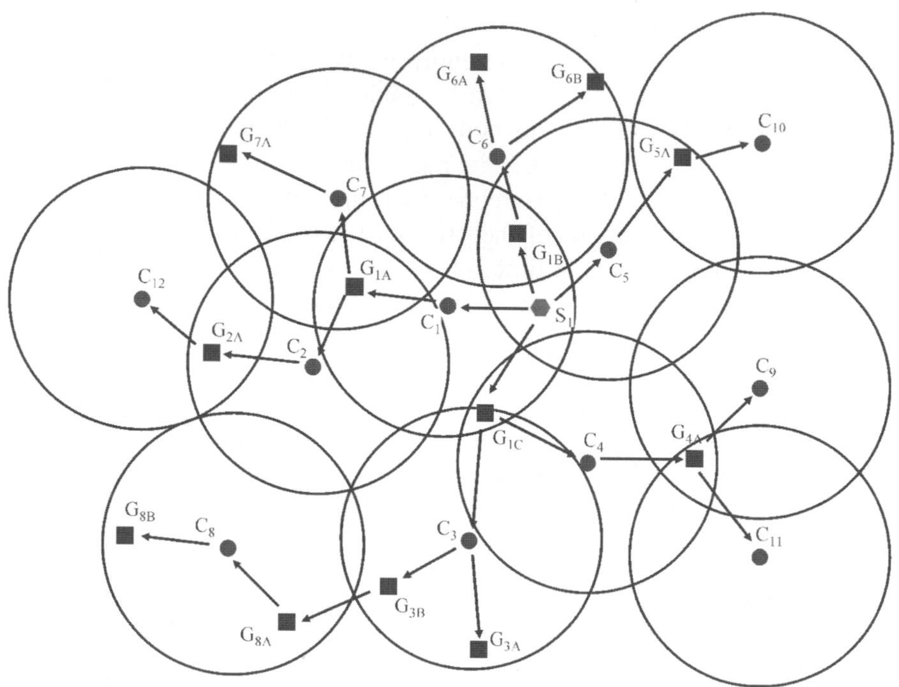


*Figure 10.1.* Illustration of NB-TRACE broadcasting. The hexagon represents the source node; disks are clusterheads; the large circles centered at the disks represents the transmit range of the clusterheads, squares are gateways, and the arrows represent the data transmissions.

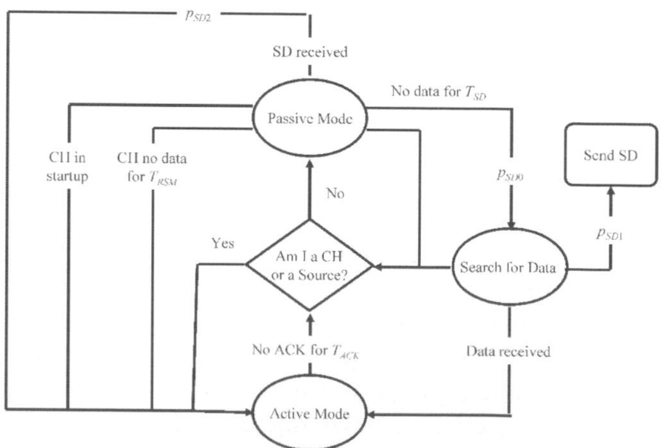


*Figure 10.2.* NB-TRACE flowchart.

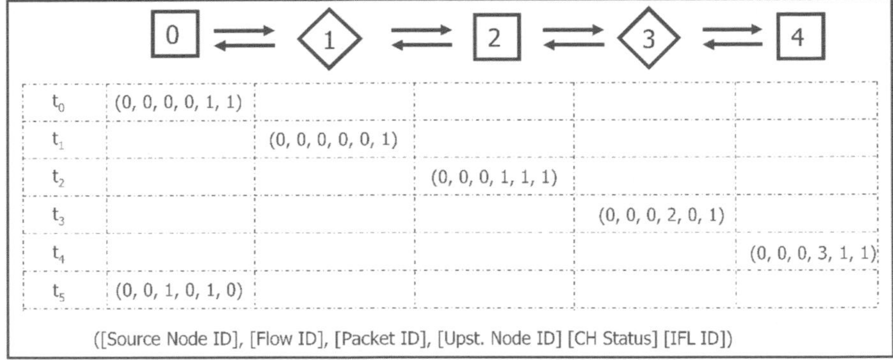

0
1
2
3
4
t0 (0, 0, 0, 0, 1, 1)
t1 (0, 0, 0, 0, 0, 1)
t2 (0, 0, 0, 1, 1, 1)
t3 (0, 0, 0, 2, 0, 1)
t4 (0, 0, 0, 3, 1, 1)
t5 (0, 0, 1, 0, 1, 0)
([Source Node ID], [Flow ID], [Packet ID], [Upst. Node ID] [CH Status] [IFL ID])


*Figure 10.3.* Illustration of IFL and IS contents. Squares and diamonds represent CHs and ordinary nodes, respectively. Node-0 is the source node. The entries below the nodes represent the contents of ([Source Node ID] [Flow ID] [Packet ID] [Upstream Node ID] [(CH Status)] [IFL ID]) fields of their IS packets (ti's represent time instants).

### 10.2.4 Pruning (PRN)

During the initial flooding, the broadcast relays are determined in a distributed fashion. Actually, a node can be in only three states: (*i*) passive, (*ii*) active, and (*iii*) ACB (activate branch). Nodes in passive mode do not relay packets, they just receive them. Nodes in the active state act as relays, because they are the only nodes that participate in broadcast data forwarding, excluding the transients where some nodes temporarily stay in the active mode. During Initial Flooding all nodes that obtained channel access switch to the active state (*i.e.*, they rebroadcast data packets). Active nodes that do not receive an ACK for $T_{ACK}$ time cease rebroadcasting and return to passive mode. Nodes need to wait for $T_{ACK}$ time to cease relaying because network dynamics may temporarily prevent a downstream node from ACKing an upstream node (*e.g.*, mobility, cluster maintenance). However, this algorithm has a vital shortcoming, which will eventually lead to the silencing of all relays. The outermost (leaf) nodes will not receive any ACKs, thus they will cease relaying, which also means that they cease ACKing the upstream nodes. As such, sequentially all nodes will cease relaying and ACKing, which will limit the traffic to the source node only.

To solve this problem, we introduce another feature to the algorithm, which is that the CHs always retransmit, regardless of whether or not they receive an ACK. Thus, the broadcast tree formed by IFL and PRN always ends at CHs. Note that the CHs create a non-connected dominating set. Thus, if we ensure that all the CHs relay broadcast packets, then the whole network is guaranteed to be completely covered. However, this is not the optimal solution, because in some topologies some CHs have only one neighbor, which is their upstream node (*e.g.*, node-4 in Figure 10.3). Such redundant rebroadcasts deteriorate the

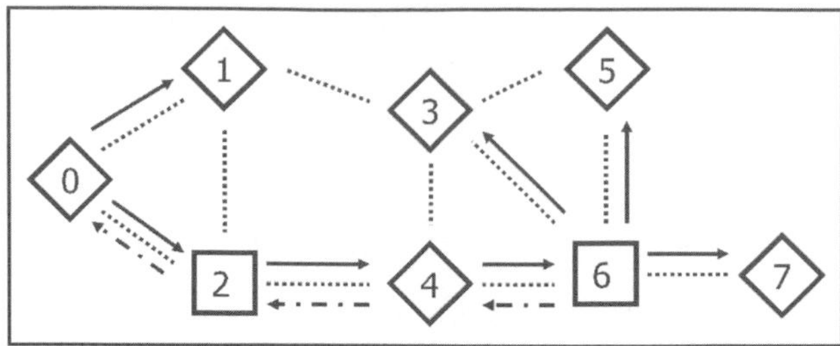


*Figure 10.4.* Illustration of IFL and PRN. Squares and diamonds represent CHs and ordinary nodes, respectively. Node-0 is the source node. Dotted lines represent the links between the nodes. Solid and dash-dotted lines represent the data and IS flows, respectively.

spatial reuse efficiency of NB-TRACE. Nevertheless, comparative evaluations of NB-TRACE with other broadcast architectures (see Section 10.3) show that overall the spatial reuse efficiency of NB-TRACE is better than other architectures.

The IFL and PRN mechanisms are illustrated in Figure 10.4. After the IFL all the nodes in the network receive the data packets and they determine whether they are receiving ACKs or not. Eventually, nodes 1, 3, 5, and 7 cease retransmitting and switch to the passive mode and nodes 0, 2, 4, and 6 stay in the active mode (*i.e.*, the broadcast tree consists of nodes 0, 2, 4, and 6 and all the other nodes are in the one-hop neighborhood of at least one tree member). Node-5 ceases rebroadcasting $T_{ACK}$ time after its first data transmission because it does not receive any ACK from any of its neighbors; however, until that time, node-5 is ACKing node-3. Node-3 ceases rebroadcasting $2T_{ACK}$ time after its first data transmission, because for the first $T_{ACK}$ time, node-3 was being ACKed by node-5. With the same token, node-1 ceases rebroadcasting $3T_{ACK}$ time after its first data transmission. Although the pruning of redundant rebroadcasts takes some time, this does not introduce any additional delay in data traffic (*i.e.*, all of the members of the broadcast tree start to relay packets as soon as they get channel access).

The first two blocks of the algorithm (IFL and PRN) are sufficient to create a broadcast tree for a static network. However, for a dynamic (mobile) network, we need extra blocks in the algorithm, because due to mobility the broadcast tree will be broken in time. The simplest solution would be to repeat the IF block periodically, so that the broken links will be repaired (actually recreated) periodically. Although this algorithm is simple, it would deteriorate the overall bandwidth efficiency of the network. The quest for more efficient compensation mechanisms lead us to design three maintenance procedures.

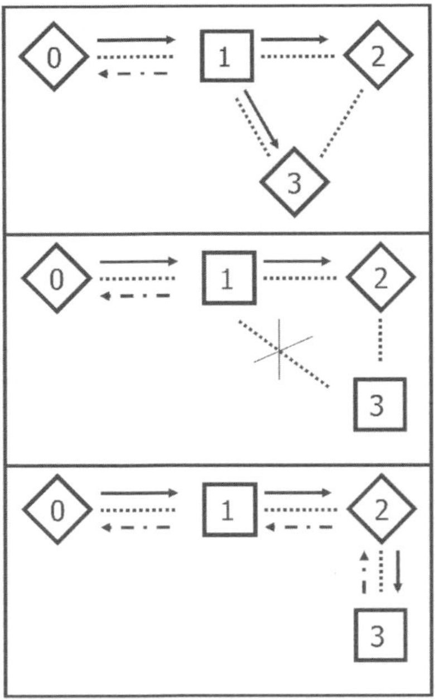


*Figure 10.5.* Illustration of the RPB mechanism.

### 10.2.5 Repair Branch (RPB)

One of the major effects of node mobility on NB-TRACE is the resignation of existing CHs and the appearance of new CHs (*i.e.*, when two CHs enter each others' receive range, one of them resigns, and if there are no CHs in the receive range of a node, it contends to become a CH). At the beginning of its operation as a CH, the CH stays in startup mode until it sends its header packet and announces its status with a bit included in the beacon packet. The appearance of a new CH generally is associated with the resignation of an existing CH. Whatever the actual situation, the nodes that receive a beacon packet from a CH in startup mode switch to active mode and rebroadcast the data packets they receive from their upstream neighbors until they cease to relay due to pruning. Figure 10.5 illustrates the RPB mechanism in a simple scenario. In the upper panel only node-1 is a CH and the broadcast tree consists of nodes 0 and 1. Nodes 2 and 3 receive data packets through node-1. However, due to the movement of node-3 (center panel), node-1 is out of the reach of node-3, thus, node-3 becomes a CH. Upon receiving the beacon of node-3, which indicates that it is in the startup mode, node-2, which was in the passive mode, switches to the active mode, thus, node-3 starts to receive data packets from node-2 (lower panel).

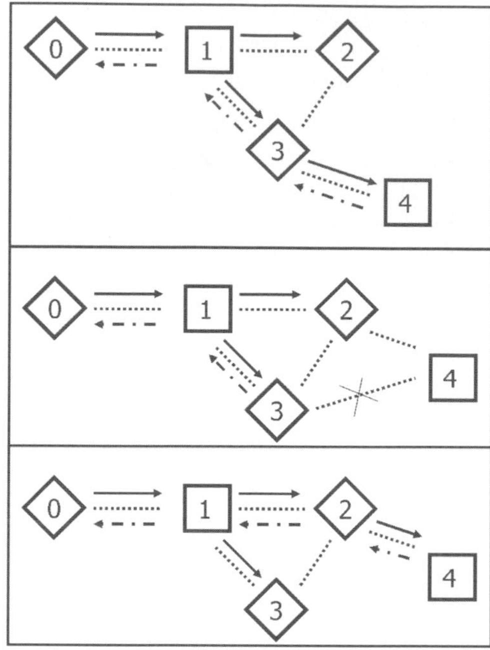


*Figure 10.6.* Illustration of the CRB mechanism.

Although RPB significantly improves the system performance in combating node mobility, it cannot completely fix the broken tree problem. For example, a CH could just move away from its only upstream neighbor, which creates a broken tree. This problem (and other similar situations) cannot be handled by RPB. Thus, we introduce CRB, which, in conjunction with RPB, almost completely alleviates the tree breakage problem.

### 10.2.6 Create Branch (CRB)

One of the basic principles of the NB-TRACE algorithm is that all the CHs should be rebroadcasting. If an ordinary node detects any of the CHs in its receive range is inactive for $T_{CRB}$ time, then it switches to active mode and starts to rebroadcast data. As in the RPB case, redundant relays will be pruned in $T_{ACK}$ time. The CRB mechanism is illustrated in an example scenario in Figure 10.6. Node-4, which is a CH, receives data through node-3 (upper panel). Due to mobility node-4 moves away from node-3 and the link between node-3 and node-4 is broken. However, node-4 enters into the receive range of node-2 (center panel). Upon detecting an inactive CH (node-4) in its receive range for $T_{CRB}$ time, node-2 switches to the active mode and node-4 starts to receive data from its new upstream node, which is node-2 (lower panel).

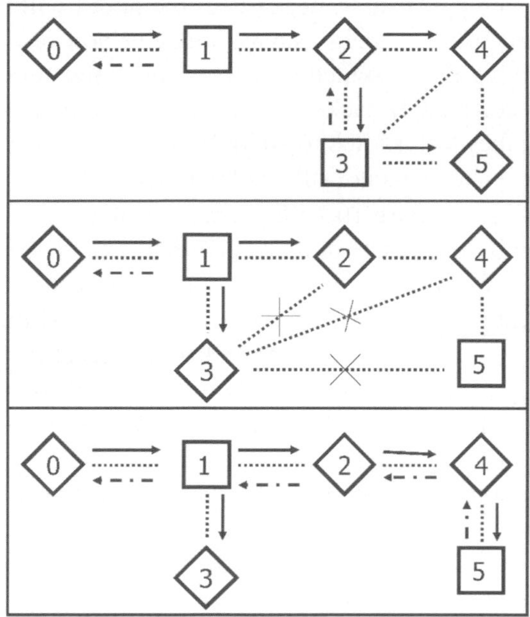


*Figure 10.7.* Illustration of the ACB mechanism.

The first four building blocks (IFL, PRN, RPB, and CRB) create an almost complete broadcasting algorithm capable of handling mobility. However, in some scenarios none of the aforementioned blocks can repair broken links. Figure 10.7 illustrates such a scenario. The upper panel of Figure 10.7 illustrates a network with a complete broadcast tree, where all the nodes can receive the broadcast packets. After some time, due to the mobility of nodes 3 and 5, node-3 gives up being a CH and node-5 becomes a CH (center panel). However, the potential upstream node of node-5, which is node-4, cannot relay data packets to node-5 by using the CRB mechanism, because the potential upstream node of node-4, which is node-2, is in the passive mode and does not supply node-4 with data packets. The ACB block comes into play at this point to fix this problem.

### 10.2.7 Activate Branch (ACB)

An ordinary node that does not receive any data packets for $T_{ACB}$ time switches to ACB mode, and sends an ACB packet with probability $p_{ACB}$. The underlying MH-TRACE MAC does not have a structure that can be used for this purpose, thus we modified MH-TRACE to be able to send ACB packets without actually affecting any major building blocks of MH-TRACE. ACB packets are transmitted by using the IS slots, because all the nodes will be listening to the IS slots regardless of their energy saving mode. Upon reception of an ACB packet, the receiving nodes switch to active mode, and start to relay data. If the nodes that receive ACB packets do not have data to send, they are either in ACB mode or they will switch to ACB mode. Upon receiving the first data packet, the nodes in ACB mode will switch to active mode.

The lower panel of Figure 10.7 illustrates the ACB mechanism. Upon detecting the inactivity of the CH in its receive range (node-5), node-4 switches to active mode in $T_{CRB}$ time, which does not help to fix the broken link in this situation. After $T_{ACB}$ time node-4 switches to ACB mode and sends its ACB packet to node-2. Upon receiving and ACB packet from node-4, node-2 switches to active mode and starts to relay the data packets it receives from its upstream node (node-1) to node-4, and then node-4 switches to active mode and relays the packets to node-5.

The main functionality of ACB is to activate an inactive distributed gateway formed by two ordinary nodes (*i.e.*, nodes 2 and 4 in Figure 10.7). Once the distributed gateway is activated, then the flow of data packets, possibly from multiple flows and/or from multiple sources, can reach the leaf CH (*i.e.*, node-5), hence the nodes in the CH's one-hop neighborhood. Thus, ACB is not for actually searching for any particular broadcast flow from any particular source node, rather it is for connecting the disconnected section of the broadcast tree to the rest of the broadcast tree.

### 10.2.8 Multiple Packet Drop Thresholds

Utilizing a single packet drop threshold throughout the network is not a good strategy, because of the fact that the source node does not drop packets until the packet delay exceeds the packet drop threshold. Due to the network dynamics, packet delay is accumulated in time. When packets are transmitted by the source node at the verge of being dropped, these packets cannot be relayed and are dropped by the neighbors of the source node. The remedy for this problem is to use a two-level packet drop threshold scheme. Thus, in NB-TRACE two packet drop thresholds are utilized. A large packet drop threshold, $T_{drop}$, dictated by the application is used throughout the whole network, and a smaller packet drop threshold, $T_{drop-source}$, is used only at the source node so that the packets that would not be relayed due to large delays are automatically dropped by the source node. We set $T_{drop-source}$ to be equal to the packet generation period, $T_{PG}$ because we want to keep $T_{drop-source}$ as small as possible to minimize the overall delay and we do not want to drop a packet before there is another packet ready in the queue.

## 10.3 Simulations and Analysis

### 10.3.1 Simulation Environment

We explored the QoS and energy dissipation characteristics of NB-TRACE, flooding with MH-TRACE, and flooding, gossiping, Counter-based broadcasting (CBB), and Distance-based broadcasting (DBB) (see Section 4.3.3 with IEEE 802.11 and SMAC through extensive ns-2 simulations within the traffic load and node density parameter space. All the simulations are run for 100 s and averaged over 10 runs. We used a CBR traffic generator with UDP transport agent to simulate a constant rate voice codec. We used the propagation model described in Section 2.2.1. We used two energy models in this chapter. the first model has the following parameters: $P_T = 0.60$ W, $P_R = 0.30$ W, $P_I = 0.10$ W, and $P_S = 0.01$ W. We also used the energy model presented in [102] (*i.e.*, $P_T = 2.50$ W, $P_R = 0.90$ W, $P_I = 0.11$ W, and $P_S = 0.02$ W), for comparison. Unless otherwise stated, the default energy model in this chapter is the former one. We used the same overhead for IEEE 802.11, SMAC, and MH-TRACE data packets in our simulations (*i.e.*, 10 bytes) to compare all protocols on a fair basis; therefore, all data packets are 110 bytes. MH-TRACE and NB-TRACE control packets are 10 bytes. We evaluated the performance of the broadcast architectures with a single flow at a time. Acronyms, descriptions and values of the constant parameters used in the simulations are given in Table 10.1.

We used the RWP mobility model where the node speeds are chosen from a uniform random distribution between 0.0 m/s and 5.0 m/s (see Section 2.4). The pause time is set to zero to avoid non-moving nodes throughout the simulation time. As reported in [64, 65], we observe a decrease in the average instantaneous node speed with time. Average node speed at the beginning is $2.5 \pm 0.2$ m/s; however, at the end (100 s) the average node speed decreases to $2.2 \pm 0.4$ m/s. Note that, in $x \pm y$ notation, $x$ and $y$ are the mean and standard deviation of an ensemble, respectively.

We simulated several network/MAC combinations to evaluate their performance against NB-TRACE. We have chosen IEEE 802.11, SMAC, and MH-TRACE as the MAC layers because, (*i*) the IEEE 802.11 standard is well known by the wireless community, and almost all researchers compare their algorithms with IEEE 802.11, (*ii*) SMAC is the most prominent example of a truly distributed energy aware MAC protocol based on CSMA, and (*iii*) MH-TRACE is an example of a clustering based approach and a TDMA based channel access scheme. We have chosen four network layer broadcast algorithms: flooding, gossiping, Counter-Based Broadcasting (CBB), and Distance-Based Broadcasting (DBB). Flooding and gossiping are examples of non-coordinated broadcast algorithms, whereas CBB and DBB are examples of partially coordinated broadcast algorithms. Thus, our comparisons span a wide range of algorithms for network-wide broadcasting.

*Table 10.1.* Simulation parameters.

| Acronym | Description | Value |
|---|---|---|
| $C$ | channel rate | 2 Mbps |
| $D_{Tr}$ | transmit range | 250 m |
| $D_{CS}$ | carrier Sense range | 507 m |
| $T_{drop}$ | packet drop threshold | 150 ms |
| $T_{drop-source}$ | packet drop threshold at source | 25 ms |
| $T_{RAD}$ | random assessment delay | 12.5 ms |
| $P_T$ | transmit power | 600 mW |
| $P_R$ | receive power | 300 mW |
| $P_I$ | idle power | 100 mW |
| $P_S$ | sleep power | 10 mW |
| NA | data packet overhead | 10 bytes |
| NA | control packet size | 10 bytes |
| NA | header packet size | 22 bytes |
| $T_{IFS}$ | inter-frame space duration | 16 $\mu s$ |
| $T_{ACK}$ | ACK time | $4T_{SF}$ |
| $T_{CRB}$ | CRB time | $5T_{SF}$ |
| $T_{ACB}$ | ACB time | $6T_{SF}$ |
| $p_{ACB}$ | ACB probability | 0.5 |
| $T_{SMAC}$ | SMAC sleep/active cycle period | 25 ms |
| $R_{SMAC}$ | SMAC sleep/active cycle ratio | 0.25 ms |
| $p_{GSP}$ | gossiping probability | 0.1-0.9 |
| $N_{CBB}$ | CBB maximum counter | 2-7 |
| $D_{DBB}$ | DBB minimum distance | 140 m - 240 m |

### 10.3.2 General Performance Analysis

In this subsection we present the simulation results for NB-TRACE and all the other architectures in a 1 km by 1 km network with 80 nodes. Data rate is 32 Kbps, which is realized by 100-byte payload packets with 25 ms packet generation period. We analyze the broadcast architectures independently and at the end we compare them.

MH-TRACE-based flooding average and minimum packet delivery ratios (PDRs) are both 99% (see Table 10.2). Average packet delay and delay jitter of MH-TRACE are 47.8 ms and 2.1 ms, respectively. MH-TRACE average number of retransmitting nodes per data packet (ARN) is 61.2. Note that not all of the nodes are retransmitting even though the network layer algorithm is flooding. One reason for such behavior is that the number of data slots available is less than the number of nodes in the network and thus some nodes are denied channel access.

*Table 10.2.* MH-TRACE and NB-TRACE performance.

| | MH-TRACE | NB-TRACE No Data Traffic | NB-TRACE | NB-TRACE 3B | NB-TRACE 4B $T_{drop-source} \rightarrow 150$ ms | NB-TRACE |
|---|---|---|---|---|---|---|
| $PDR_{AVG}$ | $99.6 \pm 0.2\%$ | $99.8 \pm 0.8\%$ | NA | $82.0 \pm 6.8\%$ | $99.4 \pm 0.2\%$ | $89.2 \pm 13.0$ % |
| $PDR_{MIN}$ | $99.4 \pm 0.3\%$ | $99.7 \pm 0.2\%$ | NA | $24.5 \pm 12.4\%$ | $96.9 \pm 2.5\%$ | $68.4 \pm 22.7$ % |
| ARN | $61.2 \pm 7.9\%$ | $19.2 \pm 1.5\%$ | NA | $13.7 \pm 2.3\%$ | $18.1 \pm 1.2\%$ | $18.0 \pm 3.6$ % |
| Delay (ms) | $47.8 \pm 6.6\%$ | $35.9 \pm 4.0\%$ | NA | $40.0 \pm 5.3\%$ | $35.1 \pm 4.8\%$ | $93.6 \pm 15.0$ % |
| Jitter (ms) | $2.1 \pm 0.1\%$ | $2.1 \pm 0.1\%$ | NA | $2.1 \pm 0.1\%$ | $2.1 \pm 0.1\%$ | $2.4 \pm 0.3$ % |
| Total ED (mJ/s) | $55.8 \pm 0.7\%$ | $44.5 \pm 1.1\%$ | $31.0 \pm 0.2\%$ | $40.5 \pm 1.2\%$ | $43.5 \pm 0.6\%$ | $43.4 \pm 1.6$ % |
| Transmit ED (mJ/s) | $8.8 \pm 0.3\%$ | $3.6 \pm 0.4\%$ | $0.3 \pm 0.0\%$ | $2.3 \pm 0.3\%$ | $3.2 \pm 0.2\%$ | $3.1 \pm 0.5$ % |
| Receive ED (mJ/s) | $13.5 \pm 0.2\%$ | $9.6 \pm 0.4\%$ | $1.8 \pm 0.1\%$ | $7.7 \pm 0.6\%$ | $9.3 \pm 0.2\%$ | $9.0 \pm 0.9$ % |
| CS ED (mJ/s) | $14.2 \pm 0.9\%$ | $9.1 \pm 0.1\%$ | $3.8 \pm 0.2\%$ | $7.3 \pm 0.6\%$ | $8.3 \pm 0.4\%$ | $8.9 \pm 0.6$ % |
| Idle ED (mJ/s) | $11.4 \pm 0.3\%$ | $14.3 \pm 0.4\%$ | $16.9 \pm 0.1\%$ | $15.3 \pm 0.2\%$ | $14.7 \pm 0.1\%$ | $14.4 \pm 0.3$ % |
| Sleep ED (mJ/s) | $7.9 \pm 0.0\%$ | $8.0 \pm 0.0\%$ | $8.1 \pm 0.0\%$ | $8.0 \pm 0.0\%$ | $8.0 \pm 0.0\%$ | $8.0 \pm 0.0$ % |

Average and minimum PDRs of NB-TRACE are both above 99%. NB-TRACE average packet delay, 32.9 ms, is 31% less than MH-TRACE average delay due to the network layer coordination in NB-TRACE. RMS jitter, 2.1 ms, is less than 10% of the average delay.

ARN of NB-TRACE is $19.2 \pm 1.5$ retransmissions per generated packet, which is approximately one third of the ARN of MH-TRACE. Due to the reduction in the number of packet transmissions, NB-TRACE dissipates 20% less energy than MH-TRACE. Before analyzing the energy dissipation of NB-TRACE with data traffic, we present an analysis with zero data traffic (*i.e.*, no data packets are generated). NB-TRACE per node energy dissipation with no data traffic is presented in Table 10.2. Transmit energy is dissipated on the control packet (beacon, CA, header) transmissions by the CHs; receive and carrier sense energy is dissipated for the reception of control packets; idle energy is dissipated during the IS slots by all nodes and during the contention slots by the CHs. Energy dissipation in the transient periods (*i.e.*, startup, network maintenance) also affects all of the energy dissipation terms. On the average, 81.4% of the total time is spent in the sleep mode and 16.7% of the total time is spent in the idle mode. Only 2.8% of the total time is spent in transmit, receive, and carrier sense modes; however, 19.4% of the total energy dissipation is due to these modes, because of the higher power level of transmit and receive/carrier sense, when compared to idle and sleep modes.

When we compare NB-TRACE energy dissipations with and without data transmissions, we see that sleep mode energy dissipations and time spent in the sleep mode are almost the same for both cases because the only difference when data is present is that the relay nodes switch to transmit mode once in each superframe and all the nodes switch to receive mode once to receive a single copy of a new data packet; thus average sleep time shows only a small decrease (*e.g.*, from $8.1 \pm 0.0$ mJ/s to $8.0 \pm 0.0$ mJ/s). Data packet transmissions constitute 82.4% of the transmit energy dissipation, IS packet transmissions and the all the other control packet transmissions follows with 7.5% and 10.1%, respectively. NB-TRACE transmit energy dissipation is 41% of the transmit energy dissipation of MH-TRACE flooding due to the reduction in the ARN. NB-TRACE with data dissipates more energy on receive and carrier sensing and less in the idle mode when compared to the zero traffic case because the IS slots are not inactive anymore. For the same reason, MH-TRACE-based flooding energy dissipation in the receive and carrier sense modes are higher and idle mode is lower than NB-TRACE.

To observe the effects of the various blocks of NB-TRACE, we ran simulations with several subsets of the five blocks (see Table 10.2). NB-TRACE 3B and NB-TRACE 4B use the first three (*i.e.*, IFL, PRN, and RPB) and four blocks (*i.e.*, IFL, PRN, RPB, and CRB), respectively. NB-TRACE 3B average and minimum PDR are 82.0% and 24.5%, respectively. The main reason for such low PDRs is the lack of block 4 (CRB). NB-TRACE 4B has approximately the same average PDR as the full NB-TRACE (99.4%); however, the minimum PDR of NB-TRACE 4B, $96.9 \pm 2.5\%$, is slightly lower than full NB-TRACE, which is due to the inactive distributed gateway situation. Total energy dissipation per node per second of NB-TRACE 4B, 43.5 mJ/s, is 2.3% less than that of full NB-TRACE, 44.5 mJ/s, due to the ACB block, which quantifies the extra energy dissipation caused by the ACB block.

To illustrate the validity of our concern about the single packet drop threshold, we ran a simulation where we set $T_{drop-source}$ to $T_{drop}$ (150 ms). The results, listed in Table 10.2, show that the average and minimum PDRs of NB-TRACE with $T_{drop-source} = 150$ ms are 89.2% and 68.4%, respectively, and the average packet delay is 93.6 ms. The reason for such behavior is that a significant portion of the data packets transmitted by the source node have delays close to $T_{drop}$, and after one or two hops they are dropped.

Table 10.3 presents the performance comparisons and rankings of broadcast architectures with the best performance configurations, which are determined through simulations. Random assessment delay, $T_{RAD}$, is chosen, approximately, as half the packet generation period, $T_{PG}$, so that a broadcasting node will not contend against another node for the next packet within its two-hop radius. SMAC sleep/active cycle period, $T_{SMAC}$, is matched to $T_{PG}$, so that data packets are not accumulated in the queue (*i.e.*, if $T_{SMAC}$ is longer than $T_{PG}$, then a node needs to send two packets during one active period and packet delay increases). SMAC sleep/active ratio, $R_{SMAC}$, is chosen as 0.25, so that SMAC can save energy without decreasing system performance (*i.e.*, higher $R_{SMAC}$ results in lower effective bandwidth and higher energy savings). Gossiping with IEEE 802.11 and SMAC PDRs (average and minimum) reach their maximums at $p_{GSP} = 0.7$ and $p_{GSP} = 0.9$, respectively. Higher and lower $p_{GSP}$'s result in lower PDRs due to the higher collision rates and incomplete coverage, respectively. CBB with IEEE 802.11 and SMAC have the highest PDRs with $N_{CBB} = 3$ and $N_{CBB} = 2$, respectively. We simulated DBB with 20 m $D_{DBB}$ steps to pick the best configurations. DBB with IEEE 802.11 and SMAC PDRs are maximized with $D_{DBB} = 200$ m and $D_{DBB} = 220$ m, respectively.

We use short acronyms (see Table 10.4) for the architectures to be able to fit them into a single table. In terms of PDR, jitter, ARN, and energy efficiency, NB-TRACE is either the best or as good as all the other architectures. The ARN and energy saving performance of NB-TRACE is especially superior to the other schemes. However, NB-TRACE delay is not among the best. In fact, NB-TRACE packet delay is ranked in the second half of all the architectures. Nevertheless, for the scenario we considered in this subsection, NB-TRACE delay does not create a vital problem since it is still below the packet drop

*Table 10.3.* General performance comparisons (NB: NB-TRACE, MH: MH-TRACE-based flooding, CI: CBB with IEEE 802.11 and $N_{CBB} = 3$, DI: DBB with IEEE 802.11 and $D_{DBB} = 200$ m, GI: Gossiping with IEEE 802.11 and $p_{GSP} = 0.7$, FI: Flooding with IEEE 802.11, CS: CBB with SMAC and $N_{CBB} = 2$, DS: DBB with SMAC and $D_{DBB} = 220$ m, GS: Gossiping with SMAC and $p_{GSP} = 0.9$, FS: Flooding with SMAC).

| Rank | $PDR_{AVG}$ / $PDR_{MIN}$ | Delay (ms) | Jitter (ms) | ARN | Energy-1 (mJ/s) | Energy-1 normalized | Energy-2 (mJ/s) | Energy-2 normalized |
|---|---|---|---|---|---|---|---|---|
| 1 | NB<br>$99.8 \pm 0.1\%$ / $99.7 \pm 0.2\%$ | CI<br>$11.5 \pm 1.1$ | NB<br>$2.1 \pm 0.1$ | NB<br>$19.2 \pm 1.5$ | NB<br>$44.5 \pm 1.1$ | NB<br>1.0 | NB<br>$102.5 \pm 5.5$ | NB<br>1.0 |
| 2 | CI<br>$99.7 \pm 0.2\%$ / $99.3 \pm 0.3\%$ | CS<br>$12.3 \pm 0.5$ | MH<br>$2.1 \pm 0.1$ | CS<br>$23.2 \pm 0.2$ | MH<br>$55.8 \pm 0.7$ | MH<br>1.3 | MH<br>$148.0 \pm 4.7$ | MH<br>1.5 |
| 3 | MH<br>$99.6 \pm 0.2\%$ / $99.4 \pm 0.3\%$ | GI<br>$17.0 \pm 0.7$ | CI<br>$6.0 \pm 0.7$ | CI<br>$31.2 \pm 2.5$ | CS<br>$131.2 \pm 0.2$ | CS<br>2.9 | CS<br>$298.3 \pm 0.7$ | CS<br>2.9 |
| 4 | DI<br>$99.2 \pm 0.1\%$ / $99.1 \pm 0.1\%$ | DI<br>$24.8 \pm 0.8$ | CS<br>$6.7 \pm 0.5$ | GI<br>$55.6 \pm 0.1$ | CI<br>$171.5 \pm 0.1$ | CI<br>3.9 | CI<br>$392.7 \pm 0.3$ | CI<br>3.9 |
| 5 | GI<br>$97.6 \pm 0.3\%$ / $94.8 \pm 0.3\%$ | FI<br>$28.8 \pm 0.8$ | GI<br>$6.8 \pm 0.8$ | DS<br>$58.5 \pm 0.0$ | DS<br>$204.4 \pm 0.1$ | DS<br>4.6 | DS<br>$587.1 \pm 0.7$ | DS<br>5.7 |
| 6 | CS<br>$98.5 \pm 0.1\%$ / $92.0 \pm 0.5\%$ | NB<br>$35.9 \pm 4.0$ | DI<br>$9.6 \pm 0.5$ | MH<br>$61.2 \pm 7.9$ | GS<br>$205.0 \pm 0.1$ | GS<br>4.6 | GS<br>$590.8 \pm 0.4$ | GS<br>5.8 |
| 7 | DS<br>$92.5 \pm 0.1\%$ / $85.5 \pm 0.2\%$ | MH<br>$47.8 \pm 6.6$ | GS<br>$26.0 \pm 0.8$ | GS<br>$66.3 \pm 0.1$ | FS<br>$205.2 \pm 1.1$ | FS<br>4.6 | FS<br>$592.5 \pm 1.3$ | FS<br>5.8 |
| 8 | GS<br>$92.2 \pm 0.2\%$ / $84.9 \pm 0.2\%$ | DS<br>$90.0 \pm 0.8$ | FI<br>$26.5 \pm 0.5$ | DI<br>$66.9 \pm 0.8$ | GI<br>$222.0 \pm 0.3$ | GI<br>5.0 | GI<br>$594.2 \pm 1.4$ | GI<br>5.8 |
| 9 | FS<br>$91.5 \pm 0.2\%$ / $83.1 \pm 0.2\%$ | GS<br>$95.0 \pm 0.8$ | FS<br>$27.0 \pm 0.8$ | FS<br>$71.6 \pm 0.5$ | DI<br>$229.6 \pm 0.1$ | DI<br>5.2 | DI<br>$625.0 \pm 0.6$ | DI<br>6.1 |
| 10 | FI<br>$91.0 \pm 0.7\%$ / $90.0 \pm 0.8\%$ | FS<br>$96.0 \pm 0.2$ | DS<br>$30.0 \pm 0.8$ | FI<br>$73.7 \pm 0.3$ | FI<br>$243.9 \pm 0.4$ | FI<br>5.5 | FI<br>$682.1 \pm 1.9$ | FI<br>6.7 |

*Table 10.4.* Acronyms and descriptions for the broadcast architectures presented in Table 10.3.

| Acronym | Description |
|---|---|
| NB | NB-TRACE |
| MH | MH-TRACE-based flooding |
| CI | CBB with IEEE 802.11 and $N_{CBB} = 3$ |
| DI | DBB with IEEE 802.11 and $D_{DBB} = 200$ m |
| GI | Gossiping with IEEE 802.11 and $p_{GSP} = 0.7$ |
| FI | Flooding with IEEE 802.11 |
| CS | CBB with SMAC and $N_{CBB} = 2$ |
| DS | DBB with SMAC and $D_{DBB} = 220$ m |
| GS | Gossiping with SMAC and $p_{GSP} = 0.9$ |
| FS | Flooding with SMAC |

threshold. CBB with IEEE 802.11 is the architecture with the lowest packet delay. Furthermore, it is the second best architecture, overall.

NB-TRACE, MH-TRACE-based flooding, CBB with IEEE 802.11, and DBB with IEEE 802.11 are the broadcast architectures that achieved higher than 99.0% minimum PDR. CBB with SMAC and Gossiping with IEEE 802.11 average PDR's are above 95%; however, their minimum PDRs are below 95%, which are below acceptable limits. All the other architectures with flooding and SMAC produced average and minimum PDR's below 95%. Thus, regardless of the other metrics only the top four architectures (NB-TRACE, MH-TRACE-based flooding, and CBB and DBB with IEEE 802.11) are within acceptable QoS limits.

The top two minimum delay broadcast architectures are CBB with IEEE 802.11 and SMAC, which have average packet delays of 11.5 ms and 12.3 ms, respectively. The second group is formed by Gossiping with IEEE 802.11 and DBB with IEEE 802.11, which have average packet delays of 17.0 ms and 24.8 ms, respectively. The third tier consists of Flooding with IEEE 802.11, NB-TRACE, and MH-TRACE, which have packet delays of 28.8 ms, 35.9 ms, and 47.8 ms. The largest delay group is all other SMAC architectures: DBB with SMAC, Gossiping with SMAC, and Flooding with SMAC, which have packet delays of 90.0 ms, 95.0 ms, and 96.0 ms, respectively.

NB-TRACE and MH-TRACE are the top two in jitter rankings with 2.1 ms RMS jitter. The second group is formed by CBB with IEEE 802.11 and SMAC, Gossiping with IEEE 802.11, and DBB with IEEE 802.11, ranging from 6.0 ms to 9.4 ms. The highest jitter is observed for Gossiping with SMAC, Flooding with IEEE 802.11, Flooding with SMAC, and DBB with SMAC, ranging from 26.0 ms to 30.0 ms.

NB-TRACE and CBB with SMAC ARNs are the lowest among all the architectures, at 19.2 and 23.2, respectively. CBB with IEEE 802.11 ARN is the third lowest, 31.2, and all the other ARNs are distributed between 55.6 and 73.7. As expected, Flooding with IEEE 802.11 has the highest ARN, 73.7.

In Table 10.3, we present energy dissipation results with two different energy models, described in Section 10.3.1. In both energy models, NB-TRACE and MH-TRACE are the two lowest energy dissipating architectures. CBB with SMAC and CBB with IEEE 802.11 energy dissipations are in between the first group formed by NB-TRACE and MH-TRACE and the highest energy dissipation group formed by the rest of the architectures. The relative energy efficiency of NB-TRACE is better with the second energy model than the first energy model, which is our default model, yet energy dissipation rankings are not affected from the energy model change. Energy dissipation distributions of the broadcast architectures are presented in Figure 10.8. All of the distributions are approximately centered around their average energy dissipation values, thus, average energy dissipation is a sufficient statistical metric to evaluate the energy dissipation characteristics of these architectures. Furthermore, NB-TRACE distribution is well separated from the non-MH-TRACE-based distributions. For example, the highest energy dissipating node of NB-TRACE will live at least twice as long as the lowest energy dissipating node of CBB with SMAC, which is the lowest energy dissipating non-MH-TRACE-based architecture.

Having completed our analysis for this particular set of parameters (*i.e.*, data rate and node density) we now investigate the effects of varying the data rate on NB-TRACE. For the rest of the simulations we compare the performance of NB-TRACE with CBB with IEEE 802.11 and DBB with IEEE 802.11 only, since these architectures are the only other architectures that produced acceptable QoS except MH-TRACE-based flooding. Since network conditions get harsher with increasing node density and/or data rate, the already unacceptably low performance of the other architectures will deteriorate further.

### 10.3.3 Varying the Data Rate

In this subsection we explore the effects of varying the data rate on the protocol performance. The data rate is varied by changing the size of the data packets by keeping the packet generation period constant. NB-TRACE parameters (*e.g.*, number of frames within a superframe, $N_F$, and number of data slots per frame, $N_D$) are reconfigured with the changing data packet sizes (see Table 10.5). The number of frames and the number of data slots per frame along with other parameters (*e.g.*, the number of contention slots) are adjusted to keep the superframe time approximately 25.0 ms.

NB-TRACE, CBB, and DBB performances as a function of data rate are presented in Table 10.6 for an 80 node 1 km by 1 km network. NB-TRACE average and minimum PDR stays above $95\%$ for all data rates. The drop in PDR

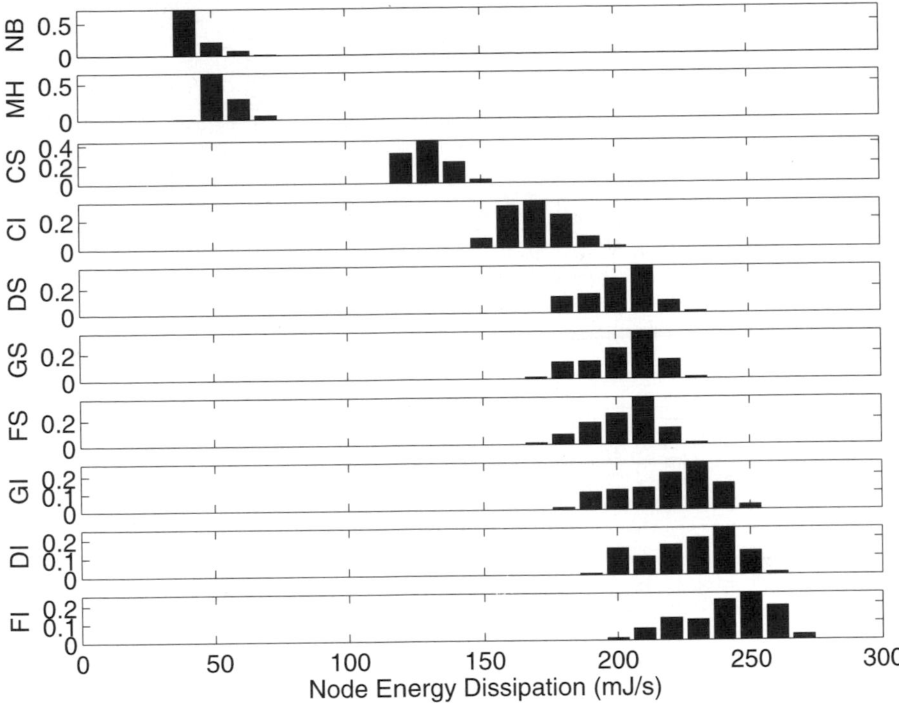


*Figure 10.8.* Normalized histograms of node energy dissipations for different broadcast architectures.

*Table 10.5.* NB-TRACE parameters: Number of frames per superframe, $N_F$, number of data slots per frame, $N_D$, and data packet payload.

| Data rate (Kbps) | $N_F$ | $N_D$ | Payload (Bytes) |
|---|---|---|---|
| 16 | 6 | 9 | 50 |
| 32 | 6 | 6 | 100 |
| 48 | 6 | 4 | 150 |
| 64 | 6 | 3 | 200 |
| 80 | 6 | 3 | 250 |
| 96 | 6 | 2 | 300 |

with increasing data rate is due to the small number of data slots per CH, which limits the operation characteristics of NB-TRACE. For example, the number of data slots that can be utilized in response to an ACB request decreases with increasing data rate. NB-TRACE ARN stays almost constant around 19.0 for all data rates. Both delay and jitter slightly increase with increasing data rate.

*Table 10.6.* Performance of NB-TRACE, CBB, and DBB as a function of data rate.

| | Data Rate (Kbps) | 16 | 32 | 48 | 64 | 80 | 96 |
|---|---|---|---|---|---|---|---|
| NB-TRACE | $PDR_{AVG}$ | $99.9 \pm 0.1\%$ | $99.8 \pm 0.1\%$ | $99.7 \pm 0.3\%$ | $99.5 \pm 0.4\%$ | $99.4 \pm 0.5\%$ | $98.4 \pm 1.1\%$ |
| | $PDR_{MIN}$ | $99.9 \pm 0.1\%$ | $99.7 \pm 0.2\%$ | $99.4 \pm 0.3\%$ | $98.4 \pm 0.5\%$ | $96.7 \pm 0.5\%$ | $95.6 \pm 0.5\%$ |
| | ARN | $19.1 \pm 0.9$ | $19.2 \pm 1.5$ | $19.1 \pm 1.2$ | $19.0 \pm 1.3$ | $19.6 \pm 0.7$ | $18.7 \pm 0.2$ |
| | Delay (ms) | $35.6 \pm 3.5$ | $35.9 \pm 4.0$ | $36.5 \pm 4.3$ | $37.1 \pm 3.7$ | $37.0 \pm 3.8$ | $39.4 \pm 4.2$ |
| | Jitter (ms) | $2.1 \pm 0.1$ | $2.1 \pm 0.1$ | $2.1 \pm 0.1$ | $2.1 \pm 0.1$ | $2.2 \pm 0.2$ | $2.5 \pm 0.5$ |
| | Energy (ms) | $45.9 \pm 0.7$ | $44.5 \pm 1.1$ | $44.1 \pm 0.7$ | $46.1 \pm 0.6$ | $47.7 \pm 0.5$ | $50.8 \pm 0.8$ |
| CBB | $PDR_{AVG}$ | $99.9 \pm 0.1\%$ | $99.7 \pm 0.2\%$ | $99.6 \pm 0.3\%$ | $92.6 \pm 0.3\%$ | $90.2 \pm 0.4\%$ | $90.0 \pm 0.4\%$ |
| | $PDR_{MIN}$ | $99.9 \pm 0.1\%$ | $99.3 \pm 0.3\%$ | $98.3 \pm 0.3\%$ | $88.8 \pm 0.4\%$ | $87.5 \pm 0.5\%$ | $81.5 \pm 0.5\%$ |
| | ARN | $28.9 \pm 0.1$ | $31.2 \pm 0.3$ | $36.2 \pm 0.1$ | $57.2 \pm 0.2$ | $58.8 \pm 0.4$ | $59.2 \pm 0.2$ |
| | Delay (ms) | $9.9 \pm 0.2$ | $11.5 \pm 0.3$ | $12.6 \pm 0.6$ | $64.3 \pm 1.0$ | $95.7 \pm 0.5$ | $106.0 \pm 0.8$ |
| | Jitter (ms) | $6.0 \pm 0.3$ | $6.0 \pm 0.3$ | $6.0 \pm 0.5$ | $19.2 \pm 0.9$ | $23.0 \pm 1.2$ | $25.2 \pm 1.3$ |
| | Energy (ms) | $136.9 \pm 0.1$ | $171.5 \pm 0.2$ | $212.8 \pm 0.2$ | $267.4 \pm 0.1$ | $274.5 \pm 0.2$ | $276.9 \pm 0.1$ |
| DBB | $PDR_{AVG}$ | $99.9 \pm 0.1\%$ | $99.2 \pm 0.1\%$ | $98.2 \pm 0.2\%$ | $95.3 \pm 0.2\%$ | $91.3 \pm 0.4\%$ | $88.0 \pm 0.5\%$ |
| | $PDR_{MIN}$ | $99.9 \pm 0.1\%$ | $99.1 \pm 0.1\%$ | $98.0 \pm 0.2\%$ | $92.2 \pm 0.7\%$ | $84.0 \pm 0.9\%$ | $75.7 \pm 0.9\%$ |
| | ARN | $67.3 \pm 0.1$ | $66.9 \pm 0.1$ | $66.1 \pm 0.2$ | $64.0 \pm 0.2$ | $60.8 \pm 0.3$ | $58.0 \pm 0.1$ |
| | Delay (ms) | $12.1 \pm 0.1$ | $24.8 \pm 0.8$ | $61.0 \pm 0.8$ | $92.0 \pm 0.9$ | $102.3 \pm 1.1$ | $106.1 \pm 1.0$ |
| | Jitter (ms) | $5.0 \pm 0.1$ | $9.6 \pm 0.5$ | $22.7 \pm 0.5$ | $27.3 \pm 0.5$ | $28.0 \pm 0.7$ | $29.3 \pm 0.9$ |
| | Energy (ms) | $178.6 \pm 0.1$ | $229.6 \pm 0.1$ | $262.8 \pm 0.2$ | $272.0 \pm 0.1$ | $274.4 \pm 0.1$ | $278.3 \pm 0.1$ |

There are two factors affecting the energy dissipation of NB-TRACE: a decrease in the number of data slots, which also means a decrease in the number of IS slots and total IS time, and an increase in the amount of data transmitted and received with increasing data rate. Higher data traffic means a reduction in the total amount of time spent in the IS period, where nodes spend less energy when compared to a longer IS period. Remember that all of the nodes stay in active mode during the IS slots. Thus, the energy dissipated in the IS slots, which is a significant component of the energy dissipation in NB-TRACE, decreases with increasing data rate. On the other hand, energy dissipated on transmission and reception of data packets increases with the increasing data rates, because the packet length (amount of data) increases. Note that the number of data packets generated per packet generation time stays constant for all data rates, and each node receives at most one copy of each generated packet. Furthermore, the number of packet transmissions also stays almost constant (*i.e.*, ARN) for all data rates. Thus, when these two mechanisms are combined, the total per node energy dissipation decreases in the first half of the data rate space, reaching 44.1 mJ/s at 48 Kbps, and increases in the second half, reaching 50.8 mJ/s at 96 Kbps. Nevertheless, the variation of the energy dissipation lies in a narrow band.

For all data rates, the rebroadcast counter of CBB, $N_{CBB}$, is three. Higher values of $N_{CBB}$ resulted in unacceptable PDRs due to the increase in congestion with a higher number of retransmissions, whereas lower values of $N_{CBB}$ failed to create a complete set cover.

Average and minimum PDRs of CBB are lower than $95\%$ starting with 64 Kbps, reaching $90.0\%$ and $81.5\%$, respectively, at 96 Kbps data rate due to the increase in the congestion level of the network, which causes an increase in the average number of data collisions per transmission and the average number of dropped data packets per second.

The increase in ARN of CBB is also due to collisions. Nodes that cannot receive a packet due to collisions cannot increment their counter, which gives rise to the number of rebroadcasts and an increase in ARN. Actually, packet drops do not decrease the PDR of CBB; instead, if there were no packet drops, then the average PDR of CBB would be lower than its current value for higher data rates due to the reduction in the congestion by dropping packets.

CBB average delay exhibits two regimes: (*i*) low data rate regime, where CBB packet delay is a small fraction of NB-TRACE packet delay due to the low level of congestion and (*ii*) high data rate regime, where the balance is reversed (*i.e.*, NB-TRACE delay is a small fraction of CBB delay) due to the high level of congestion. CBB RMS jitter also shows a similar trend with CBB packet delay for the same reasons mentioned; however, at all data rates NB-TRACE jitter is a small fraction of CBB jitter. The increase in the CBB energy dissipation is due to the higher number of larger packet transmissions, receptions, and carrier sensing with the increasing traffic. CBB energy dissipation at 16 Kbps and 96 Kbps data rates are 3.0 and 5.5 times the energy dissipation of NB-TRACE, respectively.

For all data rates, the minimum rebroadcast distance of DBB, $D_{DBB}$, is 200 m. DBB minimum PDR drops below $95\%$ starting with 64 Kbps data rate, reaching $75.7\%$ at 96 Kbps data rate. Unlike CBB, DBB ARN drops with increasing data rate, reaching 58.0 at 96 Kbps data rate. DBB delay, jitter, and energy dissipation trends are similar to CBB.

With the current implementation, NB-TRACE parameters needs to be tuned for the data rate chosen. One way to alleviate this restriction is to keep the data slot length small and assign multiple data slots for higher data rate applications, which is similar to IEEE 802.15.3 [161] channel access. However, in a real deployment most probably all nodes are equipped with the same rate voice codecs, thus, all nodes can be tuned for the same parameters.

### 10.3.4 Varying the Node Density

Next we investigate the effects of node density on NB-TRACE and CBB. Table 10.7 presents the performance of NB-TRACE and CBB as a function of node density for a constant data rate source (48 Kbps) within a 1 km by 1 km area network. NB-TRACE average and minimum PDR stays above 95% for all node densities. The increase in ARN is due to the fact that the average number of CHs increases slightly with increasing node density. NB-TRACE average packet delay and delay jitter stay in a narrow band around 36.0 ms and 2.1 ms, respectively. Energy dissipation of NB-TRACE shows a slight decrease with node density, starting at 44.1 mJ/s for a 80 nodes/km$^2$ network and reaching 41.9 mJ/s for a 200 nodes/km$^2$ network.

CBB gives the highest PDRs with $N_{CBB} = 3$ for the results presented in this subsection, for the same reason described in the previous subsection. Similarly, $D_{DBB}$ is set to 200 m. CBB and DBB average and minimum PDRs drop below $95\%$ starting with the 120 nodes/km2 network, because of the high congestion. Average delay and jitter values of CBB and DBB show a steep increase with increasing node density. NB-TRACE delay is less than $38\%$ of CBB delay and jitter is less than $10\%$ of CBB jitter at 200 nodes/km$^2$ network.

### 10.3.5 Overall Analysis of the Simulation Results

We have presented a simulation-based comparative evaluation of NB-TRACE and other broadcast architectures. Detailed investigation of NB-TRACE revealed the relative impact of the different building blocks and design tradeoffs of the NB-TRACE architecture. Furthermore, the performance gains of NB-TRACE over MH-TRACE-based flooding in terms of spatial reuse and energy efficiency were quantified. Comparisons with other broadcast architec-

*Table 10.7.* Performance of NB-TRACE, CBB, and DBB as a function of node density.

| | Node Density (nodes/km$^2$) | 80 | 120 | 160 | 200 |
|---|---|---|---|---|---|
| NB-TRACE | PDR$_{AVG}$ | $99.7 \pm 0.3\%$ | $99.6 \pm 0.2\%$ | $99.4 \pm 0.5\%$ | $99.5 \pm 0.4\%$ |
| | PDR$_{MIN}$ | $99.4 \pm 0.3\%$ | $98.7 \pm 0.5\%$ | $97.5 \pm 1.3\%$ | $97.8 \pm 1.5\%$ |
| | ARN | $19.1 \pm 1.2$ | $23.2 \pm 0.6$ | $23.4 \pm 0.9$ | $23.7 \pm 1.4$ |
| | Delay (ms) | $36.5 \pm 4.3$ | $36.0 \pm 1.0$ | $36.5 \pm 1.2$ | $36.3 \pm 0.9$ |
| | Jitter (ms) | $2.1 \pm 0.1$ | $2.0 \pm 0.1$ | $2.0 \pm 0.1$ | $2.0 \pm 0.1$ |
| | Energy (ms) | $44.1 \pm 0.7$ | $43.3 \pm 0.5$ | $42.7 \pm 0.4$ | $41.9 \pm 0.3$ |
| CBB | PDR$_{AVG}$ | $99.6 \pm 0.3\%$ | $90.3 \pm 0.3\%$ | $82.2 \pm 1.2\%$ | $85.3 \pm 0.1\%$ |
| | PDR$_{MIN}$ | $98.3 \pm 0.3\%$ | $85.0 \pm 1.0\%$ | $73.2 \pm 1.2\%$ | $82.8 \pm 0.3\%$ |
| | ARN | $36.2 \pm 0.1$ | $60.8 \pm 1.4$ | $81.6 \pm 1.5$ | $99.9 \pm 0.2$ |
| | Delay (ms) | $12.6 \pm 0.6$ | $44.2 \pm 3.3$ | $83.0 \pm 0.7$ | $95.7 \pm 0.5$ |
| | Jitter (ms) | $6.0 \pm 0.5$ | $12.7 \pm 0.9$ | $21.9 \pm 0.7$ | $22.0 \pm 0.8$ |
| | Energy (ms) | $212.8 \pm 0.2$ | $249.2 \pm 1.3$ | $260.6 \pm 0.2$ | $262.0 \pm 0.1$ |
| DBB | PDR$_{AVG}$ | $98.2 \pm 0.2\%$ | $93.9 \pm 0.1\%$ | $91.7 \pm 0.2\%$ | $88.7 \pm 0.1\%$ |
| | PDR$_{MIN}$ | $98.0 \pm 0.2\%$ | $90.5 \pm 0.1\%$ | $82.3 \pm 0.4\%$ | $77.9 \pm 0.4\%$ |
| | ARN | $66.1 \pm 0.2$ | $89.9 \pm 0.1$ | $115.1 \pm 0.2$ | $141.4 \pm 0.1$ |
| | Delay (ms) | $61.0 \pm 0.8$ | $93.3 \pm 1.2$ | $103.3 \pm 0.5$ | $106.7 \pm 2.1$ |
| | Jitter (ms) | $22.7 \pm 0.5$ | $26.3 \pm 0.7$ | $29.1 \pm 1.0$ | $30.3 \pm 0.5$ |
| | Energy (ms) | $262.8 \pm 0.2$ | $266.9 \pm 0.1$ | $269.2 \pm 0.3$ | $270.7 \pm 0.1$ |

tures showed that NB-TRACE performance in terms of energy efficiency, jitter, PDR, and ARN is better than the other architectures and delay performance is worse than several other architectures at low node density and low data rate networks. However, with increasing node density and/or data rate networks, the relative delay performance of NB-TRACE becomes better than the other architectures due to the deterioration of the performance of these architectures and the relative stability of NB-TRACE with harsher network conditions.

## 10.4 Summary

In this chapter, we presented NB-TRACE, which is an energy-efficient network-wide voice broadcasting architecture for mobile ad hoc networks. In the NB-TRACE architecture, the network is organized into overlapping clusters through a distributed algorithm, where the clusterheads create a non-connected dominating set. Channel access is regulated through a distributed TDMA scheme maintained by the clusterheads. The first group of packets of a broadcast session is broadcasted through flooding, where each data rebroadcast is preceded by an acknowledgment to the upstream node. Nodes that do not get an acknowledgment for a predetermined time, except the clusterheads, cease to rebroadcast, which prunes the redundant retransmissions. The connected dominating set formed through this basic algorithm is broken in time due to node mobility. The network responds to the broken links through multiple mechanisms to ensure the maintenance of the connected dominating set. We compare NB-TRACE with four network layer broadcast routing algorithms (Flooding, Gossiping, Counter-based broadcasting, and Distance-based broadcasting) and three medium access control protocols (IEEE 802.11, SMAC, and MH-TRACE) through extensive ns-2 simulations. Our results show that NB-TRACE outperforms other network/MAC layer combinations in minimizing the energy dissipation and optimizing spatial reuse, while producing competitive QoS performance.

# Chapter 11

# MC-TRACE PROTOCOL ARCHITECTURE

## 11.1 Introduction

In Chapter 10 we presented NB-TRACE, which is a network-wide broadcasting architecture. Although NB-TRACE is shown to posses high spatial reuse efficiency, it is incapable of providing a selective group communication service. In other words, NB-TRACE always constructs a broadcast tree rather than a multicast tree, which does not necessarily span all of the nodes in a network. Furthermore, as it is shown in Appendix C, scalability of multicasting is better than broadcasting, provided that the multicast group size is finite and small when compared to the total number of nodes in the network. Thus, there is a need for another group communication architecture within the TRACE framework that supports multicasting.

There are many protocols for multicasting in mobile ad hoc networks [162], however, there is not a single protocol that jointly optimizes QoS, spatial reuse efficiency, and total energy dissipation. Thus, in this chapter we present our approach to jointly achieving these objectives: MultiCasting through Time Reservation using Adaptive Control for Energy efficiency (MC-TRACE) [48, 50].

MC-TRACE is a cross-layer design that incorporates network layer and medium access control (MAC) layer functionality into a single layer; thus, it is a monolithic design. While preserving the energy efficiency provided by the MAC layer (*i.e.*, MH-TRACE) in idle listening or unnecessary carrier sensing, MC-TRACE also improves the energy efficiency by minimizing the number of retransmissions as well as ensuring that nodes to not receive unnecessary data packets.

The remainder of this chapter is organized as follows. Section 11.2 describes the MC-TRACE architecture. The simulation environment and results are presented in Section 11.3. A chapter summary is presented in Section 11.4.

## 11.2 Protocol Architecture

MC-TRACE is a network architecture designed for energy-efficient voice multicasting. MC-TRACE is created though the integration of network layer multicasting with the MH-TRACE MAC protocol (see Chapter 7). We present a detailed description of MC-TRACE in the following subsections.

### 11.2.1 MC-TRACE Overview

MC-TRACE is built on the MH-TRACE architecture and is fully integrated with MH-TRACE, which makes MC-TRACE highly energy efficient. Although, MH-TRACE provides many advantageous features to MC-TRACE (*e.g.*, availability of controlled channel access, organization of the network into clusters) it also restricts the design of MC-TRACE in many ways.

There are five basic building blocks in MC-TRACE: (*i*) Initial Flooding (IFL), (*ii*) Pruning (PRN), (*iii*) Maintain Branch (MNB), (*iv*) Repair Branch (RPB), and (*v*) Create Branch (CRB). MC-TRACE creates a broadcast tree through flooding (IFL) and then prunes redundant branches of the tree using receiver-based (or multicast leaf node-based) feedback (PRN). It ensures every multicast node remains connected to the tree while minimizing redundancy and uses IS slots so nodes can keep track of their role in the tree (*e.g.*, multicast relay node) as well as the roles of their neighbors. Finally, MC-TRACE contains mechanisms for allowing broken branches of the tree to be repaired locally (MNB and RBP) and globally (CRB). The MC-TRACE architecture is designed for multiple multicast groups and it can support multiple flows within each multicast group. However, for the sake of clarity we will describe the architecture for a single multicast group with a single source and a single data flow.

### 11.2.2 Initial Flooding

The source node initiates a session by broadcasting packets to its one-hop neighbors. Nodes that receive a data packet contend for channel access, and the ones that obtain channel access retransmit the data they received. Eventually, the data packets are received by all the nodes in the network, possibly multiple times. Each retransmitting node acknowledges its upstream node by announcing the ID of its upstream node in its IS packet, which precedes its data packet transmission (see Figure 7.2 and Figure 11.1). The source node announces its own ID as its upstream node ID. Initially all retransmitting nodes announce a null ID as their downstream node ID. However, when an upstream node is acknowledged by a downstream node, the node updates its downstream node ID by the ID of this

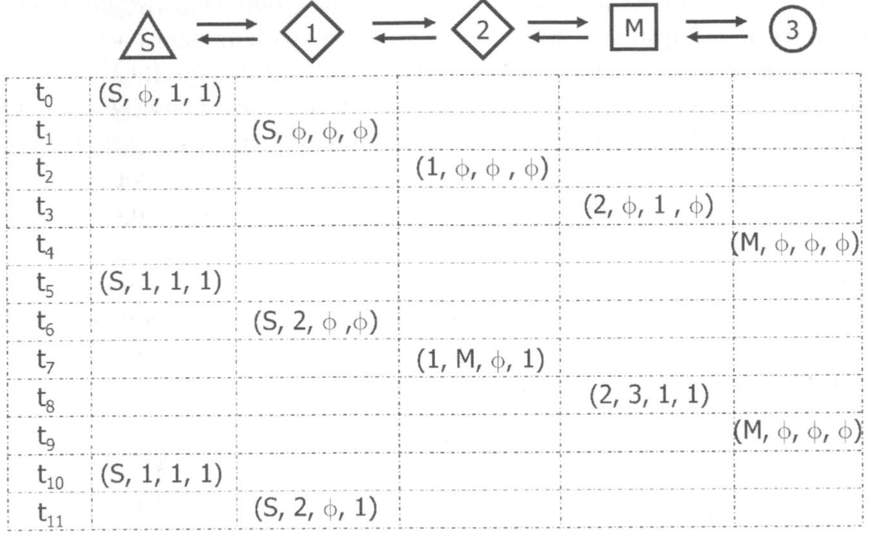

| | S | 1 | 2 | M | 3 |
|---|---|---|---|---|---|
| $t_0$ | (S, φ, 1, 1) | | | | |
| $t_1$ | | (S, φ, φ, φ) | | | |
| $t_2$ | | | (1, φ, φ , φ) | | |
| $t_3$ | | | | (2, φ, 1 , φ) | |
| $t_4$ | | | | | (M, φ, φ, φ) |
| $t_5$ | (S, 1, 1, 1) | | | | |
| $t_6$ | | (S, 2, φ ,φ) | | | |
| $t_7$ | | | (1, M, φ, 1) | | |
| $t_8$ | | | | (2, 3, 1, 1) | |
| $t_9$ | | | | | (M, φ, φ, φ) |
| $t_{10}$ | (S, 1, 1, 1) | | | | |
| $t_{11}$ | | (S, 2, φ, 1) | | | |



*Figure 11.1.* Illustration of initial flooding. Triangles, squares, diamonds, and circles represent sources, multicast group members, multicast relays, and non-relays, respectively. The entries below the nodes represent the contents of ([Upstream Node ID], [Downstream Node ID], [Multicast Group ID], [Multicast Relay Status]) fields of their IS packets (f represent null IDs and ti's represent time instants).

node. The leaf nodes (*i.e.*, nodes that do not have any downstream nodes that are acknowledging them as upstream nodes) continue to announce the null ID as their downstream node ID.

At this point, some of the nodes have multiple upstream nodes (*i.e.*, multiple nodes that have lower hop distance to the source than the current node) and downstream nodes (*i.e.*, multiple downstream nodes acknowledging the some upstream node as their upstream node). A node with multiple upstream nodes chooses the upstream node that has the least packet delay as its upstream node to be announced in its IS slot. Since a retransmitting node indicates its hop distance to the source (HDTS) in its IS packet, it is possible to choose the node with the least HDTS as the upstream node; however, our primary objective is minimizing delay rather than minimizing the multicast tree size. A node updates its own HDTS by incrementing the least HDTS it hears within $T_{HDTS1}$ time. The initial HDTS value is set to HDTS$_{MAX}$, and the HDTS value is again set to HDTS$_{MAX}$ if a node does not receive any IS or data packet for more than $T_{HDTS2}$ time, where $T_{HDTS2}$ is larger than $T_{HDTS1}$.

Multicast group member nodes indicate their status by announcing their multicast group ID in the IS packet (see Figure 11.1). Nodes that are not members of the multicast group set their multicast group ID to the null multicast group ID. If an upstream node receives an acknowledgment (ACK) from a downstream multicast group member, it marks itself as a multicast relay and announces its multicast relay status by setting the corresponding status (*i.e.*, multicast relay bit) in the IS packet. The same mechanism continues in the same way up to the source node. In other words, an upstream node that gets an ACK from a downstream multicast relay marks itself as a multicast relay. Furthermore, a multicast group member that receives an ACK from an upstream multicast relay marks itself as a multicast relay also. Multicast relay status expires if no ACK is received from any downstream (for both members and non-members of the multicast group) or upstream (only for members of the multicast group) multicast relay or multicast group member for $T_{RLY}$ time. For the sake of simplicity, we assume a link between any node pair is bidirectional at this point; however, this is not necessary for MC-TRACE to operate successfully. Initial flooding results in a highly redundant multicast tree, where most of the nodes receive the same data packet multiple times. Thus, a pruning mechanism is needed to eliminate the redundancies of the multicast tree created by the initial flooding.

### 11.2.3 Pruning

Actually initial flooding and pruning are two mechanisms working simultaneously; however, we describe these as sequential mechanisms to make them easier to understand. During the initial flooding, the multicast relays are determined in a distributed fashion. Pruning uses the multicast relays to create an efficient multicast tree. As described previously, a multicast relay node that does not receive any upstream or downstream ACK for $T_{RLY}$ time ceases to be a multicast relay (for the sake of simplicity, we assume the multicast group members are always the leaf nodes). Furthermore, a node, which is not a multicast relay also ceases to retransmit the multicast data if it does not receive an ACK from any downstream node.

Figure 11.2 illustrates the operation of the pruning mechanism. After the initial flooding all the nodes receive the data packets and they determine their upstream and downstream nodes. Multicast relays are also determined. Nodes 4, 5, and M along with S are multicast relays. However, nodes 1, 2, and 3 are not multicast relays, because there is not a multicast group member connected to that branch of the network. Node-3 will cease retransmitting the packets that it received from its upstream node-2 $T_{RLY}$ time after its first retransmission of data, because no node is acknowledging its data transmissions. However, until that time node-3 acknowledges its upstream node, which is node-2. Node-2 ceases retransmitting packets $2T_{RLY}$ times after its first data transmission. Note that node-2 acknowledges its upstream node (node-1) for $2T_{RLY}$ time. Node-1 ceases retransmitting $3T_{RLY}$ time after its first data transmission.

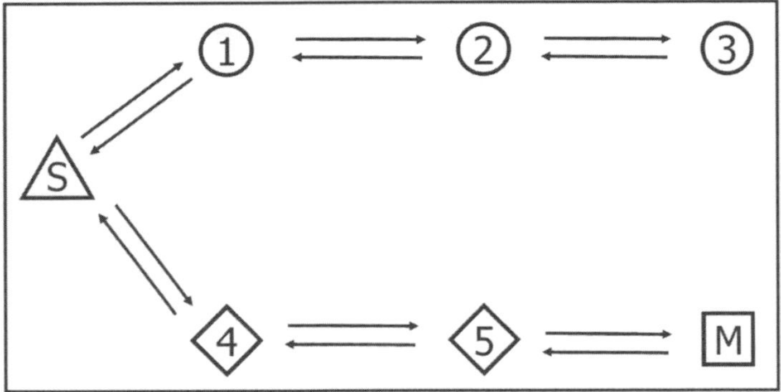


*Figure 11.2.* Illustration of pruning and multicast tree creation.

Thus, the redundant upper branch, where no multicast group members are present, is pruned.

Unlike the upper branch, the lower branch is not pruned due to the fact that the lower branch has a multicast node as the leaf node. Node-M acknowledges the upstream node (node-5) upon receiving the first data packet. Since node-5 receives an ACK form its downstream node (node-M) and also node-M indicates its multicast group membership in its IS packet, node-5 marks itself as a multicast relay and announces its status in its following IS transmission. Upon receiving that IS packet from its downstream node (node-5), node-4 marks itself as a multicast relay also. Thus, the branch of the multicast tree consisting of node-4, node-5, and node-M is created in a distributed fashion. When compared to completion of the pruning of the upper branch the completion of the creation of the lower branch is realized in much shorter time.

Although in most cases initial flooding and pruning are capable of creating an initial efficient multicast tree, they are not always capable of maintaining the multicast tree in a mobile network. Thus, the need for additional mechanisms to repair broken branches is obvious. Maintain Branch, Repair Branch, and Create Branch mechanisms are utilized to maintain the multicast tree.

### 11.2.4 Maintain Branch

Some of the multicast group members are not multicast relays. The upper panel of Figure 11.3 illustrates such a situation. Multicast node (node-M1) is a multicast relay, which is indicated by the two-way arrows; whereas node-M2 is not a multicast relay it just receives the packets from the upstream node (node-2). Hence, node-M2 does not acknowledge node-2 (node-2 is acknowledged by node-M1. Note that any node can acknowledge only one upstream and one downstream node with a single IS packet. When node-M1 moves away from node-2's transmit range and enters node-1's transmit range, it either begins to

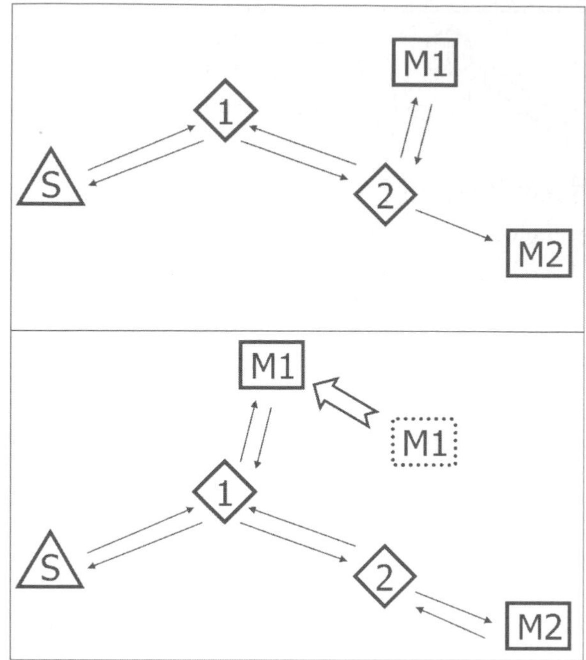


*Figure 11.3.* Illustration of the Maintain Branch Mechanism.

acknowledge node-1 as its upstream node if the transition happens in less than $T_{RLY}$ time (*i.e.*, node-M1's multicast relay status does not expire before $T_{RLY}$ time) or just receives the data packets from node-1 without acknowledging node-1 if node-M1's transition takes more than $T_{RLY}$ time. In any case, node-2 does not receive any ACK from node-M1, and starts to set its downstream node ID as the null ID. However, node-2 does not cease retransmitting data packets that it receives from its upstream node (node-1) instantly, because, a multicast relay does not resets its status for $T_{RLY}$ time and continues to retransmit data packets.

Although node-M2 does not acknowledge any node, it monitors its upstream node through IS and data packets. When the upstream node of a multicast group member node (*i.e.*, node-M2) announces null ID as its downstream node ID, the multicast node (M2) starts to acknowledge the upstream node by announcing the ID of the upstream node (node-2) as its upstream node in its IS packet. Thus, node-2 continues to be a multicast relay and node-M2 becomes a multicast relay after receiving a downstream ACK from its upstream node (node-2). Actually, the situation illustrated in Figure 13 3 is just one example for MNB mechanism. There are several other situations that can be fixed by the MNB mechanism.

The MNB mechanism does not necessarily create a new branch, yet it prevents an existing operational branch from collapse. However, just maintaining the existing multicast relays is not enough in every situation. There are situations where new relays should be incorporated to the tree.

### 11.2.5 Repair Branch

After a node marks itself as a multicast relay, it continuously monitors its upstream node to detect a possible link break between itself and its upstream multicast relay node, which manifests itself as the interruption of the data flow without any prior notification. If such a link break is detected, the downstream node uses the RPB mechanism to fix the broken link. Figure 11.4 illustrates an example of a network topology where a branch of the multicast tree is broken due to the mobility of a multicast relay and fixed later by the RPB mechanism. The upper panel of Figure 11.4 shows a multicast tree formed by the source node, node-S, multicast relay nodes, node-1 and node-2, and the multicast group node, node-M, which is a multicast relay as well. Node-3 is neither a multicast relay node nor a multicast group member; however, it receives the IS packets from node-1, node-2, and node-M (*i.e.*, node-3 is in the receive range of all three nodes). After some time, as illustrated in the lower panel of Figure 13 4, node-2 moves away from its original position and node-1 and node-2 cannot hear each other; thus, the multicast tree is broken. At this point node-2 realizes that the link is broken (*i.e.*, it does not receive data packets from its upstream node anymore) and the RPB mechanism is used to fix the broken tree. Node-2 sets its RPB bit to one in the IS packets that it sends. Upon receiving a RPB indicator, all the nodes in the receive range start to retransmit data packets as they do in the initial flooding stage. One of these nodes, which is node-3 in this scenario, replaces node-2 as a multicast relay node and the multicast tree branch is repaired.

We assumed node-3 remains in the transmit range of node-1, node-2, and node-M even after node-2 moved away from node-1's transmit range. However, even if node-3 was not in the transmit range of node-2, the tree can again be fixed. Since node-M does not receive any data packets from its upstream node (node-2), it sets its RPB bit to one and announces this in its IS packet. Upon receiving the RPB of node-M, node-3 starts to relay data packets, and upon receiving an upstream ACK from node-M, marks itself as a multicast relay.

Both MNB and RPB are limited scope maintenance algorithms (*i.e.*, they can fix mostly one-hop tree breaks). However, in a dynamic network, limited scope algorithms are not capable of completely eliminating multicast tree breaks or, in some cases, the total collapse of the multicast tree. Thus, the create branch (CRB) mechanism is needed.

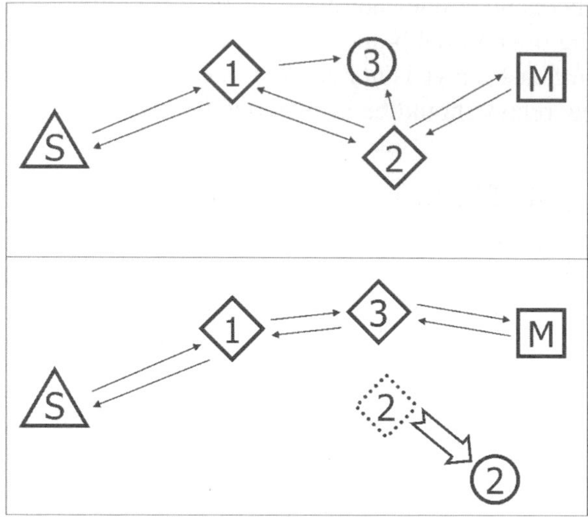


*Figure 11.4.* Illustration of the Repair Branch Mechanism.

### 11.2.6 Create Branch

It is possible that due to the dynamics of the network (*e.g.*, mobility, unequal interference) a complete branch of a multicast tree can become inactive, and the leaf multicast group member node cannot receive the data packets form the source node. Figure 11.5 illustrates a network with one active branch, composed of the nodes S, 1, 2 and M1, and one inactive branch, composed of nodes 3, 4, 5, and M2. The double arrows indicate an active link with upstream and downstream ACKs. Dash-dotted arrows indicate an inactive link. The numbers below the nodes show their HDTS, which they acquired during previous data transmissions. One situation that can create such inactivity is that the upstream ACKs of nodes 8 and M1 are colliding and node-5 cannot receive any downstream ACK. Thus, node-5 ceases to relay packets, which eventually results in silencing all the upstream nodes up to the source (*i.e.*, if node-5 does not get any downstream ACKs it ceases acknowledging its upstream node, node-4, after $T_{RLY}$ time, which results in silencing of node-4 in $2T_{RLY}$ time and node-3 in $3T_{RLY}$ time).

If a multicast group member, which is node-M2 in this scenario, detects an interruption in the data flow for $T_{CRB}$ time, it switches to Create Branch status and announces this information via a CRB packet. A CRB packet is transmitted by using one of the IS slots, which is chosen randomly. Upon receiving a CRB packet, all the nodes in the receive range of the transmitting node switch to CRB status if their own HDTS is lower than or equal to the HDTS of the sender (*e.g.*, node-5, which has an HDTS of 4, switches to CRB

BOOK ORDER

UNIVERSITY OF STRATHCLYDE 10364

| | | |
|---|---|---|
| Supplier | Dawson Books (UK) | Order |
| Order Date | 18/06/2008 | Item |
| Quantity | 1 | ISBN |
| Unit Price | 75.50 | Curre |
| Instructions | | |

| | | |
|---|---|---|
| Author | TAVLI, BULENT. | |
| Title | MOBILE AD HOC NETWORKS : ENERGY-EFFIC / BY BULENT TAVLI AND WENDI HEINZELMA | |
| Volume | | Editi |
| Format | | Publi |
| Shelf Mark | | |
| Site | ML LENDING DEWEY | |
| Fund | AEE404 | |
| Sequence | | |
| Loan Type | | |
| Quantity | 1 | |

001

95074
95074/1
9781402046322
GBP

REAL-TIME DATA COMMUNICATIONS

SPRINGER,

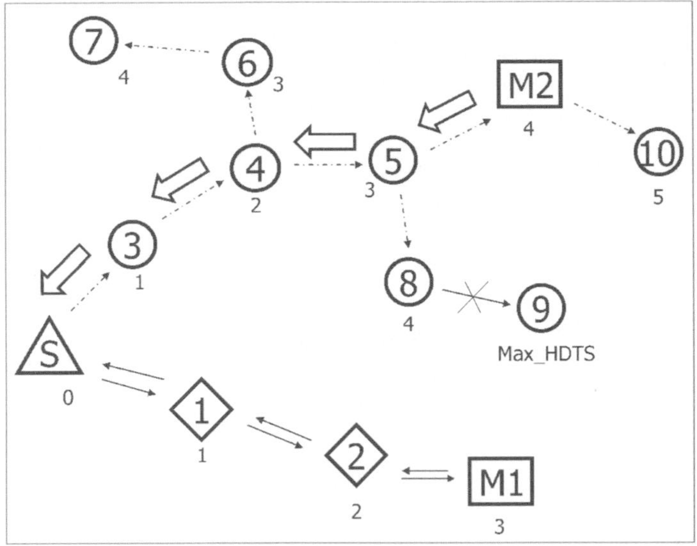


*Figure 11.5.* Illustration of the Create Branch Mechanism.

status; however, node-10, which has an HDTS of 5, does not). When a node switches to CRB mode, it starts to relay the data packets if it has data packets for the desired multicast group. If it does not have the desired data packets, it propagates the CRB request by broadcasting a CRB packet to its one-hop neighbors. This procedure continues until a node with the desired data packets is found, which is illustrated by the block arrows in Figure 11.5. After this point, the establishment of the link is similar to the initial flooding followed by pruning mechanisms. However, in this case only the nodes in CRB mode participate in data relaying. Looking at the initial collapse of the branch, we see that node-8 does not participate in CRB due to its HDTS and it does not create interference for node-M2 in this case.

## 11.2.7 Energy Efficiency

There are several mechanisms in MC-TRACE that provide energy efficiency: (*i*) nodes are in the sleep mode whenever they are not involved in data transmission or reception, which saves the energy that would be wasted in idle mode or in carrier sensing, and (*ii*) nodes can selectively choose what data to receive based on information from the IS packets, enabling the nodes to avoid receiving redundant data (*i.e.*, multiple receptions of the same packet). Note that each data packet has a unique ID, which is formed by combining the source node ID and the sequential packet ID. The sequence number need not be greater than that a few bits because data packets do not stay in the network for long due to

*Table 11.1.* Simulation parameters.

| Acronym | Description | Value |
|---|---|---|
| $C$ | channel rate | 2 Mbps |
| NA | source rate | 2 Mbps |
| NA | number of nodes | 101 |
| NA | network area | 1 km × 1 km |
| $D_{Tr}$ | transmit range | 250 m |
| $D_{CS}$ | carrier Sense range | 507 m |
| $T_{drop}$ | packet drop threshold | 150 ms |
| $T_{drop-source}$ | packet drop threshold at source | 25 ms |
| $P_T$ | transmit power | 600 mW |
| $P_R$ | receive power | 300 mW |
| $P_I$ | idle power | 100 mW |
| $P_S$ | sleep power | 10 mW |
| NA | data packet overhead | 10 bytes |
| NA | data packet payload | 100 bytes |
| NA | control packet size | 10 bytes |
| NA | header packet size | 22 bytes |
| $T_{SF}$ | superframe time | 25 ms |
| $T_{IFS}$ | inter-frame space duration | 16 $\mu s$ |
| $T_{RLY}$ | relay status expiration time | $5T_{SF}$ |
| $T_{CRB}$ | CRB time | $6T_{SF}$ |
| $T_{HDTS1}$ | HDTS decrement time | $20T_{SF}$ |
| $T_{HDTS2}$ | HDTS expiration time | $40T_{SF}$ |

the real-time requirements of the voice traffic. For example, with a packet drop threshold ($T_{drop}$) of 150 ms and packet generation period of 25 ms, there can be at most seven packets originated from a single source, simultaneously.

Although the mechanisms of MC-TRACE are fairly simple on their own, as a unified entity they create a robust architecture capable of handling complicated network dynamics, as it is shown by the simulation results.

## 11.3 Simulations and Analysis

To test the performance of MC-TRACE and to compare with IEEE 802.11 based flooding, we ran simulations using the ns-2 simulator. We used the propagation and energy models described in Section 2.2.1 and Section 5.1.1, respectively. Simulation parameters are presented in Table 11.1.

We used the random way-point mobility model for nodes moving within a 1 km by 1 km area. Node speeds are chosen from a uniform random distribution

*Table 11.2.* Performance comparison of MC-TRACE and Flooding.

| | MC-TRACE | Flooding |
|---|---|---|
| $PDR_{AVG}$ | 99% | 83% |
| $PDR_{MIN}$ | 99% | 82% |
| $Delay_{AVG}$ | 49 ms | 45 ms |
| $Delay_{MAX}$ | 78 ms | 55 ms |
| $Jitter_{AVG}$ | 2 ms | 30 ms |
| $MTS_{AVG}$ | 11 | 84 |
| $ED_{MC-AVG}$ | 50.1 mJ/s | 232.7 mJ/s |
| $ED_{MC-MAX}$ | 62.4 mJ/s | 254.3 mJ/s |
| $ED_{AN-AVG}$ | 39.4 mJ/s | 246.3 mJ/s |
| $ED_{AN-MAX}$ | 62.4 mJ/s | 272.9 mJ/s |
| $TED_{AN-AVG}$ | 50.1 mJ/s | 232.7 mJ/s |
| $RED_{AN-AVG}$ | 7.0 mJ/s | 73.3 mJ/s |
| $CSED_{AN-AVG}$ | 7.4 mJ/s | 133.6 mJ/s |
| $IED_{AN-AVG}$ | 15.5 mJ/s | 30.6 mJ/s |
| $SEP_{AN-AVG}$ | 8.0 mJ/s | 0.0 mJ/s |

between 0.0 m/s and 5.0 m/s with zero pause time (see Section 2.4). There are 100 mobile nodes in our scenario and the source node is located in the center of the network. The multicast group has five members excluding the source node.

A performance comparison of MC-TRACE and flooding is presented in Table 11.2. Both the average and the minimum packet delivery ratios (PDR) of the multicast group members for MC-TRACE are 99%, whereas those of flooding are 83% and 82%. Average PDR is the average PDR of the multicast group member nodes' PDRs. Minimum PDR is the PDR of the multicast node with minimum PDR. The difference in PDRs is due to the high congestion and consequent collisions in flooding.

Both the average and minimum data packet delays of flooding are less than those of MC-TRACE due to the restricted channel access of MC-TRACE. On the other hand, jitter obtained with flooding is 15 times the jitter obtained with MC-TRACE.

Average multicast tree size ($MTS_{AVG}$) is an appropriate metric to evaluate the spatial reuse efficiency. We determine the $MTS_{AVG}$ by dividing the total number of transmitted data packets from all nodes to the total number of transmitted data packets from the source node. MC-TRACE $MTS_{AVG}$, 11, is 13% of $MTS_{AVG}$ of flooding.

MC-TRACE average and maximum energy dissipations ($ED_{MC-AVG}$ and $ED_{MC-MAX}$) for the multicast nodes are 50.1 mJ/s and 62.4 mJ/s, respectively.

Flooding average and maximum multicast node energy dissipations are 365% and 307% more than those of MC-TRACE.

Average and minimum energy dissipations for all nodes ($ED_{AN-AVG}$ and $ED_{AN-MAX}$) are 39.4 mJ/s and 62.4 mJ/s, respectively, for MC-TRACE and 246.3 mJ/s and 272.9 mJ/s, respectively, for flooding. The difference between the transmit energy dissipation ($TED_{AN-AVG}$) is directly related with the MTS. MC-TRACE receive energy dissipation ($RED_{AN-AVG}$) is 9.5% of that of flooding due the packet discrimination (*i.e.*, redundant versions of the same packet are not received by the nodes in MC-TRACE by monitoring the IS packets). Carrier sense energy dissipation ($CSED_{AN-AVG}$) of flooding is the dominant energy dissipation term, which constitutes 54% of the total energy dissipation. Idle energy dissipation ($IED_{AN-AVG}$) of MC-TRACE is approximately half of the energy dissipation of flooding. Flooding sleep energy dissipation ($SED_{AN-AVG}$) is zero because IEEE 802.11 never goes to sleep mode.

## 11.4 Summary

In this chapter, we present MC-TRACE, which is an energy-efficient voice multicasting architecture for mobile ad hoc networks. MC-TRACE is a monolithic design, where the medium access control layer functionality and network layer functionality are performed by a single integrated layer. The basic design philosophy behind the networking part of the architecture is to establish and maintain a multicast tree within a mobile ad hoc network using broadcasting to establish the desired tree branches and pruning the redundant branches of the multicast tree based on feedback obtained from the multicast leaf nodes. Energy efficiency of the architecture is partially due to the medium access part, where the nodes can switch to sleep mode frequently; and partially due to the network layer part where the number of redundant data retransmissions and receptions are mostly eliminated. Furthermore, MC-TRACE achieves high spatial reuse efficiency by keeping the number of nodes taking part in multicasting operation minimal. We evaluated the performance of MC-TRACE through ns-2 simulations and compared with flooding. Our results show that packet delivery ratio performance, energy efficiency and spatial reuse efficiency of MC-TRACE is superior to those of flooding.

## Chapter 12

# CONCLUSIONS AND FUTURE RESEARCH DIRECTIONS

Even after a decade of intensive research and development efforts, wireless ad hoc networking is still in its infancy, with a very large design space to be explored and challenges to be overcome. Nevertheless, the fruits of these intensive and dedicated efforts have started to bloom with the appearance of currently limited but promising commercial applications [163]. Although there is a wealth of protocol architectures for wireless networking, in general, and for wireless ad hoc networking, in particular, there is still an ever increasing need for high performance wireless ad hoc network protocol architectures, as illustrated by the ever increasing number of standardization bodies and the proliferation of standards and architectures.

The work described in this book has demonstrated that a protocol architecture for mobile ad hoc networks that coordinates channel access through an explicit collective decision process based on available local information outperforms completely distributed approaches under a wide range of operating conditions in terms of QoS and energy and spatial reuse efficiency without sacrificing the practicality and scalability of the architecture. Comparative evaluations of the TRACE family of protocol architectures designed by this philosophy substantiated the performance gains achievable over other architectures in real-time data communications in mobile ad hoc networks. In Section 12.1 we present the overall conclusions of our research. Future work that builds off these contributions is addressed in Section 12.2.

## 12.1 Conclusions

In Chapter 6 we presented SH-TRACE, a TDMA-based MAC protocol for energy-efficient real-time packetized voice broadcasting in a single-hop radio network. Two features of SH-TRACE make it an energy-efficient protocol: (*i*) scheduling and (*ii*) receiver based listening cluster creation via information summarization slots. Network lifetime is maximized in SH-TRACE using dynamic controller switching and automatic backup mechanisms. Separation of the contention and data transmission is the determining factor in high throughput and stability under a very wide range of data traffic. Different QoS levels are also supported in SH-TRACE via priority levels. All of the above features are quantified through simulations and analytical models. It is shown that SH-TRACE has better energy savings and throughput performance than PRMA and IEEE 802.11.

Channel utilization of SH-TRACE and other members of the TRACE family of protocol architectures are high due to the continuous nature of an average voice burst, which spans several frames. However, channel utilization for data traffic will suffer seriously due to the non-bursty nature of data packets. Under-utilizing data slots in a Dynamic Reservation TDMA (DR-TDMA) system due to the insufficient number of successful contentions results in loss of bandwidth. As a solution to this problem, a multi-stage contention algorithm is proposed and investigated through simulations and theoretical analysis in Appendix A. The multi-stage algorithm is shown to reach the asymptotic throughput of $1/e$ and is capable of producing exactly $N$ successful contentions, on the average, in $Ne$ contention slots. The single stage algorithm cannot produce $100\%$ success, on the average, even with very large number of contention slots.

Although SH-TRACE is shown to be a high performance architecture, it is confined to operate in a fully connected single-hop network. Therefore, in Chapter 7 we presented the MH-TRACE protocol architecture, which improves and extends the SH-TRACE concepts to multi-hop networks. The most important advantages of MH-TRACE are that it provides QoS to streaming media such as voice traffic and it achieves traffic adaptive energy efficiency in a multi-hop network without using any global information except synchronization. In addition, data discrimination via receiver-based listening clusters creates an option for the application to save energy more aggressively. We used the cluster concept in such a way that: (*i*) ordinary nodes are not static members of clusters, but they choose the cluster they want to join based on the spatial and temporal characteristics of the traffic, taking into account the proximity of the cluster-heads and the availability of the data slots within the corresponding cluster; and (*ii*) each node creates its own listening cluster as if it is operating under a CSMA-type protocol. However, collisions of data packets are also minimized by means of coordination via scheduling. Thus, advantageous features of fully centralized and fully distributed networks are combined to create a hybrid and better protocol for real-time energy-efficient broadcasting in a multi-hop network.

When compared to CSMA-type broadcast protocols like 802.11, MH-TRACE has three advantages: (*i*) energy efficiency due to the use of TDMA and IS slots, which allow nodes to enter sleep mode often, (*ii*) higher throughput due to the coordinated channel access, and (*iii*) support for QoS for real-time data due to its time-frame based cyclic operation.

Minimum distance between neighboring clusterheads affects the MH-TRACE performance in terms of packet drops, collisions, and stability. Thus, in order to explore the the extent of the performance change as a function of minimum clusterhead separation we modified the MH-TRACE cluster creation and maintenance procedures, and we analyze the effects of these modified procedures in Appendix B. The conclusion we reached after analyzing the results of the simulations is that the MH-TRACE cluster creation and maintenance algorithm presented in Chapter 7 is the best alternative. Thus, the inter-clusterhead separation does not need to be treated apart from the basic cluster creation and maintenance algorithm.

Due to the dependence of the TRACE protocols on the robustness of control packets (*e.g.*, beacon, header, *etc.*), they are more vulnerable to channel errors than noncoordinated protocols (*e.g.*, IEEE 802.11), which lack such dependence due to their non-coordinated design. Thus, in Chapter 8 we investigated the impact of channel errors on the energy efficiency and QoS performance of MH-TRACE and IEEE 802.11. We developed an analytical model for the performance of MH-TRACE as a function of network area, number of nodes and BER of the channel. We presented ns-2 simulations both to demonstrate the validity of the analytical model and to show the degradation in the MAC protocols' (*i.e.*, IEEE 802.11 and MH-TRACE) performance with increasing BER. As expected, the impact of channel errors is more severe on MH-TRACE than IEEE 802.11 at extremely high BER levels due to the dependence of MH-TRACE on robust control packet traffic. Nevertheless, as the node density increases, MH-TRACE performs better than IEEE 802.11 (in terms of throughput and energy efficiency) even under very high BER levels due to its coordinated channel access mechanism.

Lessons learned from the results of Chapter 8 are not specific to MH-TRACE or IEEE 802.11. In fact, we developed our model to account for a generic schedule based coordinated MAC protocol, and the analytical model is shown to be in good agreement with the simulations, which are specific to MH-TRACE and IEEE 802.11. Thus, the major conclusion of Chapter 8 is that the energy efficiency and QoS performance of coordinated MAC protocols are superior to those of non-coordinated MAC protocols. The relatively better QoS performance of non-coordinated MAC protocols at extremely high BER levels is actually deceiving due to the fact that such a low level of QoS is not beneficial to the application layer. Finally, we point out that for higher data rates or node densities coordinated protocols are expected to perform better in terms of initial throughput due to their controlled access mechanisms.

Both SH-TRACE and MH-TRACE are designed as MAC protocols, and they do not have built-in routing mechanisms for multi-hop forwarding of data packets. In order to asses the performance of MH-TRACE in network-wide voice broadcasting and compare it with other approaches, we performed an extensive characterization of MH-TRACE and other MAC protocols in network-wide voice broadcasting through flooding in Chapter 9.

MH-TRACE-based flooding provides high energy efficiency to the nodes in the network by its coordinated medium access and data discrimination mechanisms. Especially in high data rate and/or high node density networks, the energy dissipation of MH-TRACE is less than 25% of the other schemes. Furthermore, under heavily congested networks, MH-TRACE provides satisfactory QoS to real-time data broadcasting, where the other schemes fail to fulfill the QoS requirements of the application. However, MH-TRACE packet delay performance is not as good as the other schemes, especially in mild network conditions. On the other hand, MH-TRACE packet jitter is lower than the other schemes (*e.g.*, MH-TRACE jitter is less than 10% of the IEEE 802.11 jitter at the eighth sampling path), which is as important as packet delay in multimedia applications.

IEEE 802.11-based flooding provides satisfactory QoS to real-time data broadcasting in low to medium data traffic and node densities. Furthermore, the scalability of IEEE 802.11 in mild network conditions in terms of path length is better than the other schemes due to its low packet delay. However, under heavy network conditions (high node density and data rate), IEEE 802.11 QoS performance deteriorates sharply and its scalability is also affected significantly. The energy dissipation of IEEE 802.11 is the highest among all schemes tested. Delay jitter of IEEE 802.11 is lower than SMAC and higher than MH-TRACE.

SMAC-based flooding sleep ratio shows a steep descent when the network conditions gets harsher. Furthermore, SMAC delay jitter is higher than IEEE 802.11 and MH-TRACE. SMAC can provide energy efficiency only in low node density and low data traffic networks. Yet, the scalability of SMAC is better than MH-TRACE and worse than IEEE 802.11 in such networks. However, it is not possible to employ SMAC efficiently in either high density or high data traffic networks. The main reason for such behavior is the SMAC energy saving mechanism, which reduces the energy dissipation by reducing the effective bandwidth. On the other hand, when the data rate is very low (*i.e.*, less than 8 Kbps) SMAC energy savings outperform MH-TRACE.

Data packet discrimination through information summarization is shown to be a very effective method to save energy in network-wide broadcasting through flooding, where redundant data retransmissions are unavoidable. Since each packet can be identified by its unique data packet ID, information summarization is not an ambiguous task (*i.e.*, the unique ID of each data packet is sufficient to discriminate the broadcast packets).

Utilization of multiple levels of packet drop thresholds significantly improves the broadcast performance in TDMA based schemes (*e.g.*, MH-TRACE). Furthermore, mismatches between the superframe time and the packet generation period are shown to deteriorate the PDR while improving the packet delay.

The dominant energy dissipation term for a non-energy saving protocol (*e.g.*, IEEE 802.11) in low data traffic and low node density networks is idle listening. On the other hand, in heavily congested networks, the dominant energy dissipation term is carrier sensing. Although periodic sleep/active cycling based CSMA-type medium access (*e.g.*, SMAC) can save a significant amount of energy by reducing the idle mode energy dissipation, in highly congested networks such energy saving mechanisms cannot provide satisfactory performance. Medium access control based on explicit coordination (*e.g.*, MH-TRACE) is the only option for energy savings in highly loaded networks.

The contribution of transmit energy dissipation is a minor component of the total energy dissipation in all medium access schemes. However, receive mode energy dissipation and carrier sense energy dissipation, which constitute a significant portion of the total energy dissipation, are directly related with the transmit energy dissipation. Thus, we conjecture that the impact of energy saving mechanisms targeted at minimizing the idle mode energy dissipation for mild network conditions and receive and carrier sense energy for heavy network conditions is more than the impact of the mechanisms targeted to minimize the transmit energy dissipation only, especially in broadcast scenarios.

Although MH-TRACE-based flooding achieves high energy efficiency with flooding, due to the inherent inefficiency of flooding as a broadcast routing scheme, its spatial reuse efficiency is low. Thus, in Chapter 10 we presented NB-TRACE, an energy and spatial reuse efficient network-wide broadcasting architecture. We investigated the performance of NB-TRACE and compared it with nine other broadcast architectures through extensive simulations.

There has been much research that aims to reduce the energy consumption in network wide broadcasting, however, most of this work is targeted at reducing transmit energy dissipation only with the assumption of freely available global information. On the other hand, NB-TRACE is a completely distributed algorithm, and it is targeted at reducing the total energy dissipation, which consists of not only transmit energy dissipation but receive, carrier sense, idle, and sleep energy dissipation terms as well. Although, transmit energy dissipation is the dominant energy dissipation term in some energy models, where transmit power is several orders higher than receive power; for the energy models we utilized, which are experimentally obtained from two actual radios, it is shown that transmit energy dissipation is not the dominant component of the total energy dissipation.

NB-TRACE is capable of satisfying the requirements of voice QoS (*e.g.*, PDR, delay, and jitter) under a wide range of parameters, such as data rate and node density, because of (*i*) the robustness of its distributed broadcast tree creation and maintenance algorithm, (*ii*) the explicit local coordination provided by the underlying MAC protocol, which does not create hard boundaries within the network and guarantees the availability of an underlying non-connected dominating set, (*iii*) the cross layer design, which enables the full integration of network and MAC layers, and (*iv*) distributed realization of the automatic renewal of channel access in a mobile ad hoc network and incorporating this into the tree creation and maintenance procedures.

NB-TRACE energy dissipation is much lower than the other schemes, because of (*i*) the coordinated channel access, which enables the nodes to switch to sleep mode whenever they are not involved with control or data packet traffic, (*ii*) packet discrimination, which enables nodes to avoid receiving redundant data packets, and (*iii*) comparatively lower number of rebroadcasts per generated data packet (ARN), which eliminates redundant data transmissions.

NB-TRACE packet delay is larger than some of the other broadcast architectures (CBB with IEEE 802.11 and SMAC, Gossiping with IEEE 802.11, DBB with IEEE 802.11, and Flooding with IEEE 802.11) in low node density and low data rate networks because of the restricted channel access in NB-TRACE (*i.e.*, nodes can only access the channel during their reserved data slots). However, this mechanism enables NB-TRACE to keep the average packet delay approximately constant for a wide range of parameters (*e.g.*, data rate and node density). In dense and/or high data rate networks NB-TRACE packet delay is lower than CBB and DBB with IEEE 802.11, because CBB and DBB with IEEE 802.11 packet delays exhibit a steep increase with the increasing congestion level of the network, which are related with node density and data rate.

NB-TRACE jitter is significantly lower than the other schemes except MH-TRACE based flooding. The main reason for such a low level of delay jitter is the automatic renewal of channel access (*i.e.*, once a node successfully contends for channel access, it is granted channel access automatically by the clusterhead as long as it continues to utilize its granted data slot).

NB-TRACE spatial reuse efficiency is better than the other architectures, especially in highly congested networks, because of the robustness of the channel access and the full integration of the network and MAC layers. On the other hand, other network layer broadcast algorithms, which have high spatial reuse efficiency in low traffic load networks, loose their efficiencies in high traffic load networks, because of the congestion created by the medium access control layer. For example, at high node density or high data rate networks, CBB ARN exhibits a steep increase because of the fact that the collisions due to the underlying IEEE 802.11 MAC layer prevent CBB from getting correct channel information, which gives rise to the number of retransmissions. Actually, the network layer tries to compensate for the packet collisions by increasing the retransmissions, however, the increase in the network layer rebroadcast attempts worsens the situation. Thus, the primary reason for the higher ARN of CBB in high data rate and high node density networks is the IEEE 802.11 MAC, which fails to prevent excessive collisions and causes congestion. The secondary reason is the lack of sufficient integration between the network layer (CBB) and the MAC layer (IEEE 802.11).

NB-TRACE energy savings are directly related with the energy model utilized (*i.e.*, characteristics of the radio). For example, NB-TRACE energy dissipation will be approximately the same as the other schemes for a radio that does not support a low energy sleep mode. Nevertheless, NB-TRACE continues to be an energy-efficient architecture with a radio that supports a comparatively low power sleep mode.

NB-TRACE is shown to have high spatial reuse efficiency. In other words, the broadcast capacity of NB-TRACE is higher than the broadcast capacity of other broadcast schemes. Although the asymptotic bounds on the capacity of wireless ad hoc networks for unicasting are known (*i.e.*, $O(1/\sqrt{n})$), bounds on the broadcast capacity of wireless ad hoc networks are not known. Therefore, in Appendix C, we present an upper bound on the broadcast capacity of arbitrary ad hoc wireless networks. The throughput obtainable by each node for broadcasting to all of the other nodes in a network consisting of $n$ nodes with fixed transmission ranges and $W$ bits per second channel capacity is bounded by $O(1/n)$, which is equivalent to the upper bound for per node capacity of a fully connected single-hop network. This behavior is due to the fact that routing the broadcast packets to the whole network annuls the gains from spatial reuse. Thus, the scalability of broadcasting is worse than unicasting and the scalability of multicasting is in between that of broadcasting and unicasting. Depending on the multicast group size, per node broadcast capacity of multicasting can be either $O(1/n)$, if the multicast group size is not bounded, or $O(1/\sqrt{n})$, if the multicast group size is bounded by a finite number.

Although NB-TRACE is shown to posses high spatial reuse efficiency, it is incapable of providing a selective group communication service. In other words, NB-TRACE always constructs a broadcast tree rather than a multicast tree, which is not necessarily needed. Furthermore, as it is shown in Appendix C, the scalability of multicasting is better than broadcasting, provided that the multicast group size is finite and small when compared to the total number of nodes in the network. Thus, in Chapter 11, we present MC-TRACE, which is an energy-efficient voice multicasting architecture for mobile ad hoc networks. MC-TRACE is a monolithic design, where the medium access control layer functionality and network layer functionality are performed by a single integrated layer. The basic design philosophy behind the networking part of the architecture is to establish and maintain a multicast tree within a mobile ad hoc network using broadcasting to establish the desired tree branches and pruning the redundant branches of the multicast tree based on feedback obtained from the multicast leaf nodes. Energy efficiency of the architecture is partially due to the medium access part, where the nodes can switch to sleep mode frequently;

and partially due to the network layer part where the number of redundant data retransmissions and receptions are mostly eliminated. Furthermore, MC-TRACE achieves high spatial reuse efficiency by keeping the number of nodes taking part in multicasting operation minimal. We evaluated the performance of MC-TRACE through ns-2 simulations and compared with flooding. Our results show that packet delivery ratio performance, energy efficiency and spatial reuse efficiency of MC-TRACE are superior to those of flooding. Furthermore, MC-TRACE spatial reuse efficiency is better than that of NB-TRACE for small multicast group sizes.

## 12.2 Future Research Directions

There is still much work to be done to enrich and extend the TRACE protocols. Currently, TRACE supports a single data rate because the data slot size is fixed; however, support for multiple data rate sources with different QoS requirements (*i.e.*, voice and video) necessitates modifications in the TRACE architecture. One way to overcome this problem is to introduce more degrees of freedom to the scheduling. Instead of assigning a constant duration data slot to a node, a variable duration data slot or multiple constant duration data slots can be assigned to a node depending on the QoS requirements. Furthermore, each node can get channel access from more than one clusterhead in case one clusterhead does not have enough bandwidth to meet its bandwidth demands.

Broadcasting and multicasting are two routing operations, which are well supported by TRACE (*i.e.*, NB-TRACE and MC-TRACE). Unicasting is also supported by MC-TRACE as a special case of multicasting for a multicast group size of two (*i.e.*, one source and one destination). However, for unicasting an end-to-end flow control scheme is necessary. Therefore, there is a need for a separate unicasting protocol within the TRACE framework.

Implementation of MH-TRACE on an experimental test bed is currently underway. Initial testing of a two-node MH-TRACE network has been shown to operate successfully. The nodes are created through the integration of a TI DSP chip with a PRISM IEEE 802.11 chipset. However, further prototyping for actual product development is necessary.

Information summarization has been shown to be a very effective method for energy savings in ad hoc networks. However, the information summarization methods we employed in this research are relatively simple and their scope is limited. Thus, further research on efficient information summarization is necessary. For example, for streaming video or still image traffic, efficient scene description is an effective way of summarizing information [164].

The TRACE framework can efficiently be utilized in many other application scenarios, such as sensor networking, satellite networking, and hierarchical ad hoc networking. In fact a comparison of an early version of SH-TRACE with several other architectures has confirmed the superior energy efficiency of SH-TRACE in a sensor network application for a many-to-one data transmission model [120]. Furthermore, there are several issues (*e.g.*, security and information assurance) that need to be addressed to ensure the practicality of TRACE.

Sensor and ad hoc networks are, actually, potential distributed random arrays (antennas), where the individual sensors or mobile nodes are array elements. Both in networking and sensing applications, the network can be utilized as a three dimensional active array. Synthetic aperture beamforming techniques [165–169] can be used to overcome the synchronization problems in such applications. Furthermore, through the use of spatial filtering (*i.e.*, beamforming), more efficient information fusion, detection, and classification algorithms can be created (*e.g.*, high resolution localization and tracking or very high sensitivity event classification).

# Bibliography


[1] B. Tavli. *Protocol architectures for energy-efficient real-time data communications in mobile ad hoc networks*. Ph.D. Dissertation, University of Rochester, Rochester, NY, 2005.

[2] W. Stallings. *Wireless Communications and Networks*. Prentice-Hall, 2002.

[3] Bluetooth Protocol. http://www.bluetooth.com.

[4] W. R. Crowther. A system for broadcast communications: reservation aloha. Presented at the Hawaii International Conference on System Sciences (HICCS), 1973.

[5] J.-F. Frigon, V. C. M. Leung, and H. C. B. Chan. Dynamic reservation tdma protocol for wireless atm networks. *IEEE Journal on Selected Areas in Communications*, 19:370–383, 2001.

[6] D. J. Goodman, R. Valenzuela, K. Gayliard, and B. Ramamurthi. Packet reservation multiple access for local wireless communications. *IEEE Transactions on Communications*, 37:885–890, 1989.

[7] D. J. Goodman and S. W. Wei. Efficiency of packet reservation multiple access. *IEEE Transactions on Vehicular Technology*, 40:170–176, 1991.

[8] B. O'Hara and A. Petrick. *The IEEE 802.11 Handbook: A Designer's Companion*. IEEE Press, 1999.

[9] T. M. Siep, I. C. Gifford, R. C. Braley, and R. F. Heile. Paving the way for personal area network standards: an overview of the ieee 802.15 working group for wireless personal area networks. *IEEE Personal Communications Magazine*, 7:37–43, 2000.

[10] C. R. Lin and M. Gerla. Adaptive clustering for mobile wireless networks. *IEEE Journal on Selected Areas in Communications*, 15:1265–1275, 1997.

[11] C. Zhu and M. S. Corson. A five phase reservation protocol (fprp) for mobile ad hoc networks. *ACM/Kluwer Wireless Networks Journal*, 7:371–384, 2001.

[12] Y. Yi, M. Gerla, and T. J. Kwon. Efficient flooding in ad hoc networks: a comparative performance study. In *Proceedings of the IEEE International Conference on Communications (ICC)*, pages 1059–163, 2003.

[13] A. Ephremides and T. Truong. Scheduling broadcasts in multihop radio networks. *IEEE Transactions on Communications*, 38:456–460, 1990.

[14] H. Lim and C. Kim. Multicast tree construction and flooding in wireless ad hoc networks. In *Proceedings of the ACM International Workshop on Modeling Analysis and Simulation of Wireless and Mobile Systems (MSWIM)*, pages 61–68, 2000.

[15] S. Ni, Y. Tseng, and J. Shen. The broadcast storm problem in a mobile ad hoc network. In *Proceedings of the ACM International Conference on Mobile Computing and Networking (MOBICOM)*, pages 151–162, 1999.

[16] W. Peng and X. Lu. On the reduction of broadcast redundancy in mobile ad hoc networks. In *Proceedings of the ACM International Symposium on Mobile Ad Hoc Networking and Computing (MOBIHOC)*, pages 129–130, 2000.

[17] A. Qayyum, L. Viennot, and A. Laouiti. Multipoint relaying: an efficient technique for flooding in mobile wireless networks. Technical Report 3898, The French National Institute for Research in Computer Science and Control (INRIA), 2000.

[18] J. Sucee and I. Marsic. An efficient distributed network-wide broadcast algorithm for mobile ad hoc networks. Technical Report 248, Rutgers University Center for Advanced Information Processing (CAIP), NJ, 2000.

[19] K. Tang and M. Gerla. Mac layer broadcast support in 802.11 wireless networks. In *Proceedings of the IEEE Military Communications Conference (MILCOM)*, pages 544–548, 2000.

[20] B. Williams. Network wide broadcasting protocols for mobile ad hoc networks. M.sc. dissertation, Colorado School of Mines, Golden, CO, 2002.

[21] S. Singh and C. S. Raghavendra. Pamas: power aware multi-access protocol with signaling for ad hoc networks. *ACM Computer Communication Review*, 28:5–26, 1998.

[22] W. Ye, J. Heidemann, and D. Estrin. An energy-efficient mac protocol for wireless sensor networks. In *Proceedings of the IEEE Conference on Computer Communications (INFOCOM)*, pages 1567–1576, 2002.

[23] A. Chandra, V. Gummalla, and J. O. Limb. Wireless medium access control protocols. *IEEE Communications Surveys and Tutorials*, 3:2–15, 2000.

[24] H. Zimmermann. Osi reference model - the iso model of architecture for open systems interconnection. *IEEE Transactions on Communications*, 28:425–432, 1980.

[25] A. S. Tanenbaum. *Computer Networks*. Prentice-Hall, 1996.

[26] W. B. Heinzelman. *Application-specific protocol architectures for wireless networks*. Ph.D. thesis, MIT, Cambridge, 2000.

[27] T. S. Rappaport. *Wireless Communications*. Prentice-Hall, 1996.

[28] J. Broch, D. Maltz, D. Johnson, Y. Hu, and J. Jetcheva. A performance comparison of multi-hop wireless ad hoc network routing protocols. In *Proceedings of the ACM International Conference on Mobile Computing and Networking (MOBICOM)*, pages 85–97, 1998.

[29] S. Lin and D. Costello. *Error Control Coding: Fundamentals and Applications*. Prentice-Hall, 1983.

[30] J. Hagenauer. Rate-compatible punctured convolutional codes (rcpc codes) and their applications. *IEEE Transactions on Communications*, 36:389–400, 1988.

[31] W. Ye and J. Heidemann. Medium access control in wireless sensor networks. Technical Report ISI-TR-580, USC/ISI, 2003.

[32] K. Pahlavan and A. H. Levesque. *Wireless Information Networks*. Wiley, 1995.

[33] V. Bharghavan, A. Demers, S. Shenker, and L. Zhang. Macaw: a media access protocol for wireless lans. In *Proceedings of the ACM Symposium on Communications Architectures and Protocols (SIGCOMM)*, pages 212–225, 1994.

[34] N. Abramson. Multiple access in wireless digital networks. *Proceedings of the IEEE*, 82:1360–1370, 1994.

[35] J.-C. Chen, K. M. Sivalingam, P. Agrawal, and S. Kishore. A comparison of mac protocols for wireless local networks based on battery power consumption. In *Proceedings of the IEEE Conference on Computer Communications (INFOCOM)*, volume 1, pages 150–157, 1998.

[36] K. Scott and N. Bambos. Routing and channel assigment for low power transmission in pcs. In *Proceedings of the IEEE International Conference on Universal Personal Communications*, volume 2, pages 498–502, 1996.

[37] ITU-T Recommendation G.711 - Pulse Code Modulation (PCM) of Voice Frequencies, 1988.

[38] W. B. Heinzelman, A. Chandrakasan, and H. Balakrishnan. An application-specific protocol architecture for wireless microsensor networks. *IEEE Transactions on Wireless Communications*, 1:660–670, 2002.

[39] M. Abbott and L. Peterson. Increasing network throughput by integrating protocol layers. *IEEE/ACM Transactions on Networking*, 1:600–610, 1993.

[40] D. Clark and D. Tennenhouse. Architectural considerations for a new generation of protocol. In *Proceedings of the ACM Symposium on Communications Architectures and Protocols (SIGCOMM)*, pages 200–208, 1990.

[41] B. Tavli and W. B. Heinzelman. Trace: Time reservation using adaptive control for energy efficiency. US Patent (pending), 2003.

[42] B. Tavli and W. B. Heinzelman. Mh-trace: Multi hop time reservation using adaptive control for energy efficiency. US Patent (pending), 2003.

[43] B. Tavli and W. B. Heinzelman. Mh-trace: Multi-hop time reservation using adaptive control for energy efficiency. In *Proceedings of the IEEE Military Communications Conference (MILCOM)*, pages 1292–1297, 2003.

[44] B. Tavli and W. B. Heinzelman. Trace: Time reservation using adaptive control for energy efficiency. *IEEE Journal on Selected Areas in Communications*, 21:1506–1515, 2003.

[45] B. Tavli and W. B. Heinzelman. Mh-trace: Multi-hop time reservation using adaptive control for energy efficiency. *IEEE Journal on Selected Areas in Communications*, 22:942–953, 2004.

[46] B. Tavli and W. B. Heinzelman. Pn-trace: Plain network-wide broadcasting through time reservation using adaptive control for energy efficiency. In *Proceedings of the IEEE Military Communications Conference (MILCOM)*, pages 127–133, 2004.

[47] B. Tavli and W. B. Heinzelman. Nb-trace: Network-wide broadcasting through time reservation using adaptive control for energy efficiency. In *Proceedings of the IEEE Wireless Communication and Networking Conference (WCNC)*, pages 2076–2081, 2005.

[48] B. Tavli and W. B. Heinzelman. Mc-trace: Multicasting through time reservation using adaptive control for energy efficiency. In *Proceedings of the IEEE Military Communications Conference*, 2005.

[49] B. Tavli and W. B. Heinzelman. Nb-trace: Network-wide broadcasting through time reservation using adaptive control for energy efficiency. US Patent (pending), 2005.

[50] B. Tavli and W. B. Heinzelman. Mc-trace: Multicasting through time reservation using adaptive control for energy efficiency. US Patent (pending), 2005.

[51] B. Tavli and W. B. Heinzelman. Energy and spatial reuse efficient network wide real-time data broadcasting in mobile ad hoc networking. *IEEE Transactions on Mobile Computing*, 2006. forthcoming.

[52] W. B. Heinzelman, J. Kulik, and H. Balakrishnan. Adaptive protocols for information dissemination in wireless sensor networks. In *Proceedings of the ACM International Conference on Mobile Computing and Networking (MOBICOM)*, pages 174–185, 1999.

[53] A. Kulkarni and G. Minden. Composing protocol frameworks for active wireless networks. *IEEE Communications Magazine*, 38:130–137, 2000.

[54] D. Tennenhouse and D. Wetherall. Towards an active network architecture. *ACM Computer Communications Review*, 26:5–18, 1996.

[55] H. Balakrishnan, V. Padmanabhan, S. Sesha, and R. Katz. A comparison of mechanisms for improving tcp performance over wireless links. *IEEE/ACM Transactions on Networking*, 5:756–769, 1997.

[56] J. Inouye, S. Cen, C. Pu, and J. Waipole. System support for mobile multimedia applications. In *Proceedings of the IEEE/ACM Workshop on Network and Operating System Support for Digital Audio and Video (NOSSDAV)*, pages 135–146, 1997.

[57] W. Kumwilaisak, Y. T. Hou, Q. Zhang, W. Zhu, C. C. J. Kuo, and Y. Q. Zhang. A cross-layer quality-of-service mapping architecture for video delivery in wireless networks. *IEEE Journal on Selected Areas in Communications*, 21:1685–1698, 2003.

[58] W. H. Yuen, H. Lee, and T. D. Andersen. A simple and effective cross-layer networking system for mobile ad hoc networks. In *Proceedings of the IEEE International Symposium on Personal, Indoor, and Mobile Radio Communications (PIMRC)*, pages 1952–1956, 2002.

[59] C. Intanagonwiwat, R. Govindan, D. Estrin, J. Heideman, and F. Silvia. Directed diffusion for wireless sensor networking. *IEEE/ACM Transactions on Networking*, 11:2–16, 2003.

[60] V. T. Raisinghani and S. Iyer. Cross-layer design optimizations in wireless protocol stacks. *Elsevier Computer Communications*, 27:720–724, 2004.

[61] S. Shakkottai, T. S. Rappaport, and P. C. Karlsson. Cross-layer design for wireless networks. *IEEE Communications Magazine*, 44:74–80, 2003.

[62] D. B. Johnson and D. A. Maltz. Dynamic source routing in ad hoc wireless networks. In *Mobile Computing*, chapter 5, pages 153–181. Kluwer Academic Publishers, 1996.

[63] T. Camp, J. Boleng, and V. Davies. A survey of mobility models for ad hoc network research. *Wiley Journal of Wireless Communications and Mobile Computing*, 2:483–502, 2002.

[64] J. Yoon, M. Liu, and B. Noble. Sound mobility models. In *Proceedings of the ACM International Conference on Mobile Computing and Networking (MOBICOM)*, pages 205–216, 2003.

[65] C. Bettstetter, G. Resta, and P. Santi. The node distribution of the random waypoint mobility model for wireless ad hoc networks. *IEEE Transactions on Mobile Computing*, 2:257–269, 2003.

[66] X. Hong, M. Gerla, G. Pei, and C.-C. Chiang. A group mobility model for ad hoc wireless networks. In *Proceedings of the ACM International Workshop on Modeling Analysis and Simulation of Wireless and Mobile Systems (MSWIM)*, pages 53–60, 1999.

[67] L. L. Peterson and B. S. Davie. *Computer Networks*. Academic Press, 2000.

[68] D. D. Falconer, F. Adachi, and B. Gudmundson. Time division multiple access methods for wireless personal communications. *IEEE Communications Magazine*, 33:50–57, 1995.

[69] V. Garg and J. Wilkes. *Wireless and personal communication systems*. Prentice-Hall, 1996.

[70] J. Razavilar, F. Rashid-Farrokhi, and K. Liu. Software radio architecture with smart antennas: a tutorial on algorithms and complexity. *IEEE Journal on Selected Areas in Communications*, 17:662–676, 1999.

[71] G. Foschini. Layered space-time architecture for wireless communication in fading environment when using multiple antennas. *Bell Laboratories Technical Kournal*, 1:41–59, 1996.

[72] L. G. Roberts. Aloha packet system with and without slots and capture. *ACM Computer Communications Review*, 5:28–42, 1975.

[73] S. S. Lam. Packet broadcast networks - a performance analysis of the r-aloha protocol. *IEEE Transactions on Communications*, 29:596–603, 1980.

[74] V. Kanodia, C. Li, A. Sabharwal, B. Sadaghi, and E. Knightly. Distributed multi-hop scheduling and medim access with delay and throughput constraints. In *Proceedings of*

*the ACM International Conference on Mobile Computing and Networking (MOBICOM)*, pages 200–209, 2001.

[75] P. Karn. Maca - a new channel access method for packet radio. In *Proceedings of the ARRL/CRRL Amateur Radio 9th Computer Networking Conference*, pages 134–140, 1990.

[76] R. Rozovsky and P. R. Kumar. Seedex: a mac protocol for ad hoc networks. In *Proceedings of the ACM International Symposium on Mobile Ad Hoc Networking and Computing (MOBIHOC)*, pages 167–75, 2001.

[77] HomeRF Protocol. http://www.homerf.com.

[78] C. E. Perkins and P. Bhagwat. Highly dynamic destination sequenced distance vector routing (dsdv) for mobile computers. *ACM Computer Communications Review*, 24:234–244, 1994.

[79] P. Mohapatra, J. Li, and C. Gui. Qos in mobile ad hoc networks. *IEEE Wireless Communications Magazine*, 10:44–52, 2003.

[80] D. Bertsekas and R. Gallager. *Data Networks*. Prentice-Hall, 1992.

[81] C. H. Chow. On multicast path finding algorithms. In *Proceedings of the IEEE Conference on Computer Communications (INFOCOM)*, pages 1274–1283, 1991.

[82] V. P. Kompella, J. C. Pasquale, and G. C. Polyzos. Multicast routing for multimedia communication. *IEEE/ACM Transactions on Networking*, 1:286–292, 1993.

[83] C. W. Wu and Y. C. Tay. Amris: a multicast protocol for ad hoc wireless networks. In *Proceedings of the IEEE Military Communications Conference (MILCOM)*, pages 25–29, 1999.

[84] S. J. Lee, M. Gerla, and C. C. Chiang. On-demand multicast routing protocol. In *Proceedings of the IEEE Wireless Communications and Networking Conference (WCNC)*, pages 1298–1304, 1999.

[85] C. Ho, K. Obraczka, G. Tsudik, and K. Viswanath. Flooding for reliable multicast in multi-hop ad hoc networks. In *Proceedings of the ACM International Workshop on Discrete Algorithms and Methods for Mobile Computing and Communications (DIALM)*, pages 64–71, 1999.

[86] Y.-C. Tseng, S.-Y. Ni, Y.-S. Chen, and J.-P. Sheu. The broadcast storm problem in a mobile ad hoc network. *ACM/Kluwer Wireless Networks Journal*, 8:153–167, 2002.

[87] B. Williams and T. Camp. Comparison of broadcasting techniques for mobile ad hoc networks. In *Proceedings of the ACM International Symposium on Mobile Ad Hoc Networking and Computing (MOBIHOC)*, pages 194–205, 2002.

[88] B. An and S. Papavassiliou. A mobility-based clustering approach to support mobility management and multicast routing in mobile ad-hoc wireless networks. *International Journal of Network Management*, 11:387–395, 2001.

[89] D. J. Baker and A. Ephremides. The architectural organization of a mobile radio network via a distributed algorithm. *IEEE Transactions on Communications*, 29:1694–1701, 1981.

[90] S. Basagni. Distributed clustering for ad hoc networks. In *Proceedings of the International Symposium on Parallel Architectures, Algorithms, and Networks (ISPAN)*, pages 310–315, 1999.

[91] S. Basagni. Distributed and mobility adaptive clustering for multimedia support in multihop wireless networks. In *Proceedings of the IEEE Vehicular Technology Conference (VTC)*, pages 889–893, 1999.

[92] C.-R. Dow, J.-H. Lin, A.-F. Hwang, and Y.-W. Wang. An efficient distributed clustering scheme for ad-hoc wireless networks. *IEICE Transactions on Communications*, E85-B:1561–1571, 2002.

[93] K. Xu, X. Hong, and M. Gerla. An ad hoc network with mobile backbones. In *Proceedings of the IEEE International Conference on Communications (ICC)*, pages 3138–3143, 2002.

[94] K. Xu, K. Tang, R. Bagrodia, M. Gerla, and M. Bereschinsky. Adaptive bandwidth management and qos provisioning in large scale ad hoc networks. In *Proceedings of the IEEE Military Communications Conference (MILCOM)*, pages 1018–1023, 2003.

[95] L. Williams and L. Emergy. Near trem digital radio - a first look. In *Proceedings of the Tactical Communications Conference*, pages 423–425, 1996.

[96] T. Kwon and M. Gerla. Clustering with power control. In *Proceedings of the IEEE Military Communications Conference (MILCOM)*, pages 1424–1428, 1999.

[97] A. McDonald and T. Znati. A mobility based framework for adaptive clustering in wireless ad hoc networks. *IEEE Journal on Selected Areas in Communications*, 17:1466–1486, 1999.

[98] R. Ramanathan and M. Steenstrup. Hierarchically organized multihop mobile wireless networks for quality of service support. *ACM/Baltzer Mobile Networks and Applications Journal*, 3:101–119, 1998.

[99] M. Stemm and R. Katz. Measuring and reducing energy consumption of network interfaces in handheld devices. *IEICE Transactions on Fundamentals of Electronics, Communications and Computer Sciences*, 80:1125–1131, 1997.

[100] A. Smailagic, D. P. Siewiorek, and M. Ettus. System design of low-energy wearable computers with wireless networking. In *Proceedings of the IEEE Computer Society Workshop on VLSI*, pages 25–29, 2001.

[101] V. Raghunathan, C. Schurgers, S. Park, and M. B. Srivastava. Energy-aware wireless microsensor networks. *IEEE Signal Processing Magazine*, 19:40–50, 2002.

[102] J.-P. Ebert, S. Aier, G. Kofahl, A. Becker, B. Burns, and A. Wolisz. Measurement and simulation of the energy consumption of a wlan interface. Technical Report TKN-02-010, Technical University of Berlin, Telecommunication Network Group, 2002.

[103] J. Kotwicki. An analysis of energy efficient voice over ip communications in wireless networks. M.Sc. Dissertation, Case Western Reserve University, Cleveland, 2004.

[104] L. M. Feeney and M. Nilsson. Investigating the energy consumption of a wireless network interface in ad hoc networking environment. In *Proceedings of the IEEE Conference on Computer Communications (INFOCOM)*, pages 1548–1557, 2001.

[105] J. C. Cano and P. Manzioni. Reducing energy consumption in a clustered manet using the intracluster data-dissemination protocol (icdp). In *Proceedings of the IEEE Euromicro Workshop on Parallel and Network-based Processing (EUROMICRO-PDP)*, pages 411–418, 2002.

[106] B. Chen, K. Jamieson, H. Balakrishnan, and R. Morris. Span: an energy-efficient coordination algorithm for topology maintenance in ad hoc wireless networks. *ACM/Kluwer Wireless Networks Journal*, 8:481–494, 2002.

[107] D. Qiao, S. Choi, A. Jain, and K. G. Shin. Miser: an optimal low-energy transmission strategy for ieee 802.11a/h. In *Proceedings of the ACM International Conference on Mobile Computing and Networking (MOBICOM)*, pages 161–175, 2003.

[108] V. Rodoplu and T. Meng. Minimum energy mobile wireless networks. *IEEE Journal on Selected Areas in Communications*, 17:1333–1344, 1999.

[109] C. E. Jones, K. M. Sivalingam, P. Agrawal, and J. C. Chen. A survey of energy efficient network protocols for wireless networks. *ACM/Kluwer Wireless Networks Journal*, 7:443–458, 2001.

[110] T. Cooklev. *Wireless Communication Standarts*. IEEE Press, 2004.

[111] I. F. Akyildiz, W. Su, Y. Sankarasubramaniam, and E. Cayirci. A survey on sensor networks. *IEEE Communications Magazine*, 40:102–114, 2002.

[112] R. Gandhi, S. Parthasarathy, and A. Mishra. Minimizing broadcast latency and redundancy in ad hoc networks. In *Proceedings of the ACM International Symposium on Mobile Ad Hoc Networking and Computing (MOBIHOC)*, pages 222–232, 2003.

[113] I. Chlamtac, M. Conti, and J. J. N. Liu. Mobile ad hoc networking: imperatives and challenges. *Elsevier Ad Hoc Networks Journal*, 1:13–64, 2003.

[114] F. Li and I. Nikolaidis. On minimum-energy broadcasting in all-wireless networks. In *Proceedings of the IEEE Local Computer Networks Conference (LCN)*, pages 193–202, 2001.

[115] A. F. F. Clementi, P. Crescenzi, P. Peuna, G. Rossi, and P. Vocca. On the complexity of computing minimum energy consumption broadcast subgraphs. In *Proceedings of the International Symposium on Theoretical Aspects of Computer Science (STACS)*, pages 121–131, 2001.

[116] M. Cagalj, J. P. Hubaux, and C. Enz. Minimum-energy broadcast in all-wireless networks: Np-completeness and distribution issues. In *Proceedings of the ACM International Conference on Mobile Computing and Networking (MOBICOM)*, pages 172–182, 2002.

[117] D. Li, X. Jia, and H. Liu. Energy efficient broadcast routing in static ad hoc wireless networks. *IEEE Transactions on Mobile Computing*, 3:144–151, 2004.

[118] P.-J. Wan, G. Calinescu, X.-Y. Li, and O. Frieder. Minimum energy broadcast routing in static ad hoc wireless networks. In *Proceedings of the IEEE Conference on Computer Communications (INFOCOM)*, pages 1162–1171, 2001.

[119] J. E. Wieselthier, G. D. Nguyen, and A. Epheremides. On the construction of energy-efficient broadcast and multicast trees in wireless networks. In *Proceedings of the IEEE Conference on Computer Communications (INFOCOM)*, pages 585–594, 2000.

[120] P. Chen, O'dea B, and E. Callaway. Energy efficient system design with optimum transmission range for wireless ad hoc networks. In *Proceedings of the IEEE International Conference on Communications (ICC)*, volume 2, pages 945–952, 2002.

[121] Y. Chen, E. G. Sirer, and S. B. Wicker. On selection of optimal transmission power for ad hoc networks. In *Proceedings of the Hawaii International Conference on System Sciences (HICSS)*, volume 2, page 300, 2003.

[122] C. S. R. Murthy and B. S. Manoj. *Ad Hoc Wireless Networks*. Prentice Hall, 2004.

[123] J. Janssen, D. D. Vleeschauwer, G. H. Petit, R. Windey, and J. M. Leroy. Delay bounds for voice over ip calls transported over satellite access links. *ACM/Baltzer Mobile Networks and Applications Journal*, 7:79–89, 2002.

[124] E. D. Jensen, C. D. Locks, and H. Tokuda. A time-driven scheduling model for real systems. In *Proceedings of the IEEE Real Time Systems Symposium*, pages 112–122, 1985.

[125] K. Crisler and M. Needham. Throughput analysis of reservation aloha multiple access. *IEE Electronics Letters*, 31:87–89, 1995.

[126] S. Tasaka. Stability and performance of the r-aloha packet broadcast system. *IEEE Transactions on Computers*, 32:717–726, 1983.

[127] M. Budagavi, W. Heinzelman, J. Webb, and R. Talluri. Wireless mpeg-4 video communication on dsp chips. *IEEE Signal Processing Magazine*, 17:36–53, 2000.

[128] Network simulator (ns-2). http://www.isi.edu/nsnam/ns.

[129] G. Hoblos, M. Staroswlecki, and A. Aitouche. Optimal design of fault tolerant sensor networks. In *Proceedings of the IEEE International Conference on Control Applications (CCA)*, pages 467–472, 2000.

[130] T. C. Hou and T. J. Tsai. An access based clustering protocol for multihop wireless ad hoc networks. *IEEE Journal on Selected Areas in Communications*, 19:1201–1210, 2001.

[131] J. Mannermaa, K. Kalliomaki, T. Mansten, and S. Turunen. Timing performance of various gps receivers. In *Proceedings of the IEEE International Frequency Control Symposium (FCS)*, pages 287–290, 1999.

[132] F. van Diggelen. Indoor gps theory and implementation. In *Proceedings of the IEEE Position, Location, and Navigation Symposium (PLANS)*, pages 240–247, 2002.

[133] J. Elson, L. Girod, and D. Estrin. Fine-grained network time synchronization using reference broadcasts. In *Proceedings of the Symposium on Operating Systems Design and Implementation (OSDI)*, pages 147–163, 2002.

[134] L. Huang and T.-H. Lai. On the scalability of ieee 802.11 ad hoc networks. In *Proceedings of the ACM International Symposium on Mobile Ad Hoc Networking and Computing (MOBIHOC)*, pages 173–182, 2002.

[135] J. C. Haartsen. The bluetooth radio system. *IEEE Personal Communications Magazine*, 7:28–36, 2000.

[136] P. Lettieri and M. B. Srivastava. Adaptive frame length control for improving wireless link throughput, range, and energy efficiency. In *Proceedings of the IEEE Conference on Computer Communications (INFOCOM)*, pages 564–571, 1998.

[137] D. A. Maltz, J. Broch, and D. B. Johnson. Lessons from a full-scale multihop wireless ad hoc network testbed. *IEEE Personal Communications Magazine*, 8:8–15, 2001.

[138] D. McKenna. Mobile platform benchmarks, a methodology for evaluating mobile computing devices. Technical report, Transmeta Corporation, 2000.

[139] L. F. W. van Hoesel, T. Nieberg, H. J. Kip, and P. J. M. Havinga. Advantages of a tdma based, energy-efficient, self-organizing mac protocol for wsns. In *Proceedings of the IEEE Semiannual Vehicular Technology Conference, VTC Spring*, 2002.

[140] P. J. M. Havinga and G. J. M. Smit. Energy-efficient tdma medium access control protocol scheduling. In *Proceedings of Asian International Mobile Computing Conference*, 2000.

[141] D. Kotz, C. Newport, and C. Elliot. The mistaken axioms of wireless-network research. Technical Report TR2003-467, Dartmouth College Computer Science Department, 2003.

[142] T. Numanoglu, B. Tavli, and W. B. Heinzelman. The effects of channel errors on coordinated and non-coordinated medium access control protocols. In *Proceedings of the IEEE International Conference on Wireless and Mobile Computing, Networking, and Communications*, volume 1, pages 58–65, 2005.

[143] T. Numanoglu, B. Tavli, and W. B. Heinzelman. Analysis of the impact of channel errors on coordinated wireless channel access architectures. In *Proceedings of the IEEE Military Communications Conference (MILCOM)*, 2005.

[144] T. Numanoglu, B. Tavli, and W. B. Heinzelman. Energy efficiency and error resilience in coordinated and non-coordinated medium access control protocols. *Elsevier Computer Communications Journal*, 2006. forthcoming.

[145] E. N. Gillbert. Capacity of a burst-noise channel. *Bell System Tech. J.*, 39:1253–65, 1960.

[146] E. O. Elliott. Estimates of error rates for codes on bursty-noise channels. *Bell System Tech. J.*, 42:1977–97, 1963.

[147] J. R. Yee and E. J. Weldon. Evaluation of the performance of error correcting codes on a gilbert channel. *IEEE Transactions on Communications*, 43:2316–2323, 1995.

[148] J. L. Lemmon. Wireless link statistical bit error model. Technical Report NTIA Report 02-394, U.S. Department of Commerce, 2002.

[149] M. Zorzi and R. R. Rao. Lateness of probability of a retransmission scheme for error control on a two-state markov channel. *IEEE Transactions on Communications*, 47:1537–1548, 1999.

[150] J. McDougall and S. Miller. Sensitivity of wireless network simulations to a two-state markov model channel approximation. In *Proceedings of the IEEE Global Telecommunications Conference (GLOBECOM)*, 2003.

[151] A. Konrad, B. Zhao, A. Joseph, and R. Ludwig. A Markov-based channel model algorithm for wireless networks. *ACM/Kluwer Wireless Networks Journal*, 9:189–199, 2003.

[152] A. Gurtov and S. Floyd. Modeling wireless links for transport protocols. *ACM Computer Communication Review*, 34:85–96, 2004.

[153] E. Modiano. An adaptive algorithm for optimizing the packet size used in wireless arq protocols. *ACM/Baltzer Wireless Networks Journal*, 5:279–286, 1999.

[154] Y. Lu, Y. Zhong, and B. Bhargava. Packet loss in mobile ad hoc networks. Technical report, Purdue University, Report CSD-TR 03-009, 2003.

[155] M. Takai, J. Martin, and R. Bagrodia. Effects of wireless physical layer modeling in mobile ad hoc networks. In *Proceedings of the 2001 ACM International Symposium on Mobile Ad Hoc Networking and Computing (MOBIHOC)*, pages 87–94, 2001.

[156] P. Mohapatra, C. Gui, and J. Li. Group communications in mobile ad hoc networks. *IEEE Computer Magazine*, 37:52–59, 2004.

[157] M. Zuniga and B. Krishnamachari. Optimal transmission radius for flooding in large scale sensor networks. In *Proceedings of the IEEE International Conference on Distributed Computing Systems Workshops (ICDCSW)*, pages 697–703, 2003.

[158] H. Karl and A. Willig. A short survey of wireless sensor networks. Technical Report TKN-03-018, Technical University of Berlin, 2003.

[159] C. S. Raghavendra, K. M. Sivalingam, and T. Znati, editors. *Wireless Sensor Networks*. Kluwer Academic Publishers, 2004.

[160] W. Ye, J. Heidemann, and D. Estrin. Medium access control with coordinated adaptive sleeping for wireless sensor networks. *IEEE/ACM Transactions on Networking*, 12:493–506, 2004.

[161] IEEE Standard 802.15.3: Wireless Medium Access Control (MAC) and Physical Layer (PHY) Specifications for High Rate Wireless Personal Area Networks (WPANs), 2003.

[162] S. Athanassopoulos, I. Caragiannis, C. Kaklamania, and P. Kanellopoulos. Experimental comparison of algorithms for energy-efficient multicasting in ad hoc networks. *Lecture Notes in Computer Science*, 3158:183–196, 2004.

[163] R. Bruno, M. Conti, and E. Gregori. Mesh networks: commodity multihop ad hoc networks. *IEEE Communications Magazine*, 43:123–131, 2005.

[164] S. Ingec, B. Tavli, K. Ozbas, and G. Bozdagi. Muster: Multi-platform system for efficient retrieval from multimedia databases. In *Proceedings of the SPIE (Multimedia Storage and Archiving Systems IV)*, pages 165–171, 1999.

[165] B. Tavli. Data acquisition techniques for adaptive subaperture imaging. M.sc. dissertation, Baskent University, Ankara, Turkey, 1998.

[166] B. Tavli and M. Karaman. An efficient motion estimation technique for ultrasonic subaperture imaging. In *Proceedings of the IEEE Engineering in Medicine and Biology Symposium*, volume 20, pages 816–819, 1998.

[167] M. Karaman and B. Tavli. Motion estimation using selective signal redundancy for ultrasonic subaperture imaging. In *Proceedings of the IEEE Ultrasound Symposium*, pages 1615–1618, 1998.

[168] B. Tavli and M. Karaman. Correlation processing for correction of phase distortions in subaperture imaging. *IEEE Transactions on Ultrasound, Ferroelectrics, and Frequency Control*, 46:1477–1488, 1999.

[169] M. Karaman and B. Tavli. Efficient ultrasonic synthetic aperture imaging. *IEE Electronics Letters*, 35:1319–1320, 1999.

[170] Y. Abdalla, D. Kivanc, and H. Liu. Prma with reservation subframe protocol for multimedia services in mobile communication networks. In *Proceedings of the IEEE Vehicular Technology Conference (VTC)*, pages 3538–3542, 2001.

[171] I. F. Akyildiz and J. McNair. Medium access control protocols for multimedia traffic in wireless networks. *IEEE Network Magazine*, 13:39–47, 1999.

[172] E. Callaway, P. Gorday, and L. Hester. Home networking with ieee 802.15.4: a developing standard for low-rate wireless personal area networks. *IEEE Communications Magazine*, 40:70–77, 2002.

[173] M. Kawagishi, S. Sampei, and N. Morinaga. A novel reservation tdma based multiple access scheme using adaptive modulation for multimedia wireless communication systems. In *Proceedings of the IEEE Vehicular Technology Conference (VTC)*, pages 112–116, 1998.

[174] D. Raychaudhuri and N. D. Wilson. Atm-based transport architecture for multi-services wireless personal communication networks. *IEEE Journal on Selected Areas in Communications*, 12:1401–1414, 1997.

[175] N. D. Wilson, R. Ganesh, K. Joseph, and D. Raychaudhuri. Packet cdma versus dynamic tdma for multiple access in an integrated voice/data pcn. *IEEE Journal on Selected Areas in Communications*, 11:870–884, 1993.

[176] R. L. Rivest. Network control by bayesian broadcast. *IEEE Transactions on Information Theory*, 33:323–328, 1987.

[177] A. Al-Amir and M. Gurcan. Arrival rate estimation algorithm for single group slotted aloha systems. *IEEE Communications Letters*, 5:387–389, 2001.

[178] B. Bing. Stabilization of the randomized slotted aloha protocol without the use of channel feedback information. *IEEE Communications Letters*, 4:246–251, 2000.

[179] M. Ivanovich, M. Zukerman, and F. Cameron. A study of deadlock models for a multiservice medium access protocol employing a slotted aloha signaling channel. *IEEE/ACM Transactions on Networking*, 8:800–811, 2000.

[180] P. Gupta and P. R. Kumar. The capacity of wireless networks. *IEEE Transactions on Information Theory*, 46:388–404, 2000.

[181] J. Li, C. Blake, D. S. J. De Couto, H. I. Lee, and R. Morris. Capacity of ad hoc wireless networks. In *Proceedings of the ACM International Conference on Mobile Computing and Networking (MOBICOM)*, pages 61–69, 2001.

[182] U. C. Kozat and L. Tassiulas. Throughput capacity of random ad hoc networks with infrastructure support. In *Proceedings of the ACM International Conference on Mobile Computing and Networking (MOBICOM)*, pages 55–65, 2003.

# Appendix A
# Multi-Stage Contention with Feedback

In a certain type of TDMA-based MAC protocol, sometimes referred to as Dynamic Reservation TDMA (DR-TDMA), time is organized around time frames, where contention for channel access and contention-free data transmission take place in contention sub-frames and data sub-frames, respectively [170–175]. The TRACE family of protocol architectures [1, 41–51] also fall into this category. The contention sub-frame consists of mini-slots, where nodes contend for channel access in the data sub-frame. For streaming data sources, such as voice, it is better for nodes to keep their data slots once they contend successfully until the end of a data burst. However, for asynchronous data transmission, data slot reservation does not result in throughput efficiency. Hence nodes should contend for channel access continuously [6].

Maximum channel utilization for a DR-TDMA protocol with $N$ data slots can be achieved if $N$ contending nodes can make successful data slot requests in the contention sub-frame. In order to guarantee $N$ successful contentions in $M$ contention slots using S-ALOHA in the contention sub-frame, $M$ should be very large. Hence, on the average, data slots will inevitably be underutilized with a single-stage S-ALOHA contention methodology that utilizes a feasible number of contention slots. However, by using a multi-stage contention strategy with feedback information at the beginning of each stage, it is possible to achieve $N$ guaranteed successful contentions in shorter time than a single-stage S-ALOHA system with very high probability.

In [176] a pseudo-Bayesian broadcast algorithm, which maximizes the channel utilization for S-ALOHA channels, is presented. In that algorithm, a node transmits its packet with a probability updated by the ternary feedback (*i.e.*, success, idle, collision) from the transmission history of the network. In [177], a recursive arrival rate estimation algorithm, which is used to adjust the system parameters to optimize the S-ALOHA system, is presented. Several other algorithms to optimize the throughput and stability of S-ALOHA based medium access control systems are proposed in the literature [176–178]. All of the existing algorithms are designed for the maximum throughput of an S-ALOHA system where each slot is used to transmit a data packet. In MAC designs where S-ALOHA is used as a contention mechanism to reserve data slots, optimization of contention sub-frame length is not addressed.

A generic DR-TDMA system is presented in section A.1. In section A.2, throughput analysis of single-stage S-ALOHA contention is investigated via mathematical analysis and simulation. In section A.3, the multi-stage contention problem in its general form is expressed. The optimal multi-stage contention algorithm is discussed in section A.5. A summary of this appendix is presented in section A.6.

## A.1 Generic DR-TDMA Frame Structure

We consider a generic DR-TDMA frame structure in Figure A.1, where the frame consists of a contention sub-frame, a reservation announcement slot, and a data sub-frame. There is a controller node this is responsible for contention reception and data slot reservation announcements by sending a schedule of the current frame data slot reservation list. All nodes in the network can hear each other. The number of nodes that are going to contend in the current frame, $N$, can be estimated and adjusted by using the algorithms proposed in [176–179]. Nodes transmit their request packets in the contention sub-frame; successful contentions are granted data slots through the transmission schedule.

## A.2 Single-Stage S-ALOHA Contention

A symbolic representation of single-stage S-ALOHA contention with $M$ contention slots is presented in Figure A.2, where SC is the start contention message transmitted by the controller. Nodes choose a random contention slot and send their contention requests in that slot. The expected number of total successful transmissions for a single-stage S-ALOHA system with $M$ slots and $N$ nodes is:

$$q = N\left(1 - \frac{1}{M}\right)^{N-1} \tag{A.1}$$

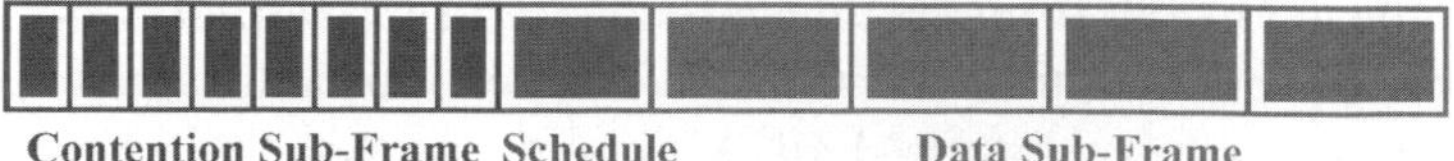


*Figure A.1.* Generic DR-TDMA frame.

Figure A.3 shows $q$ as a function of $M$ with $N = 25$. The expected value of the successful contentions is less than 25 even with $M = 1024$. Hence it is not possible to guarantee "all successful" contention with single-stage S-ALOHA even with a very large number of contention slots.

## A.3 Multi-Stage Contention

Figure A.4 shows a multi-stage S-ALOHA contention scheme. $M_i$ is the number of contention slots in the $i$'th stage of contention, $SC_i$ is the "Start Contention" packet transmitted by the controller node at the start of the $i$'th stage and contains the number of successful contentions heard in the $(i-1)$'st stage

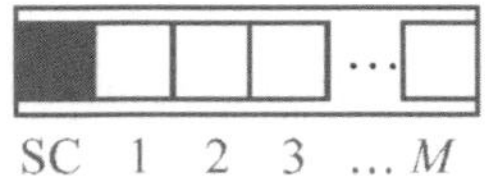


*Figure A.2.* Single-stage S-ALOHA contention.

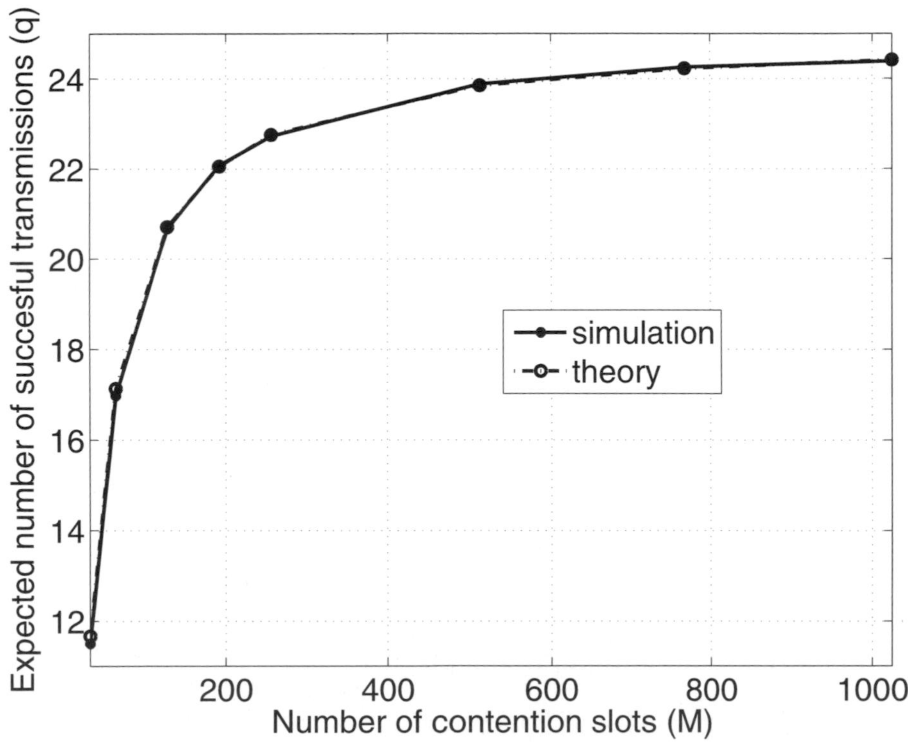


*Figure A.3.* Expected number of successful contentions vs. number of contention slots for a 25-node network ($N = 25$). Simulation results are the mean of 1000 independent runs.

and the number of contention slots in the $i$'th stage, and $K$ is the total number of contention stages. Each node will know if its contention was successful or not upon hearing the following SC, because if the number of successful contentions heard by the node and the controller is not the same, then it means the contention of the node was unsuccessful (*i.e.*, it collided with another contention packet and was not received by the controller).

## A.4 Optimal Multi-Stage Contention

We want to optimize the system parameters to minimize the time for contention, $T_C$:

$$T_C = \sum_{i=1}^{K} \{M_i T_S + T_{SC_i}\} \tag{A.2}$$

subject to the constraint that the number of successful contentions is equal to $N$, which is the total number of contending nodes. In Equation A.2, $T_S$ is the contention slot duration and $T_{SC_i}$ is the duration of the $i$'th SC packet. Since each stage is monitored independently, Equation A.2 is optimized if each contention stage is optimized independently. For $K = 1$, Equation A.2 becomes single-stage S-ALOHA contention. We want to maximize the number of successful contentions per contention slot. In order to do so, we define another quantity, the expected number of successful contentions per contention slot, $r$, as $r = q/M$. After taking the derivative of $r$ with respect to $M$ and equating to zero we find that $r$ is maximized for $M = N$. Therefore, the expression for the optimal successful number of contentions per stage is:

$$q_{opt} = N\left(1 - \frac{1}{N}\right)^{N-1} = \lim_{N\to\infty} N\left(1 - \frac{1}{N}\right)^{N-1} = \frac{N}{e} \tag{A.3}$$

which is equal to the maximum throughput of an S-ALOHA system.

Since the expected number of successful contentions is optimized for $N = M$, in each stage of the contention the number of contention slots, $M_i$, should be adjusted accordingly (*i.e.*, $M_1 = N$, $M_2 = N - N_1$, $M_3 = N - N_1 -$

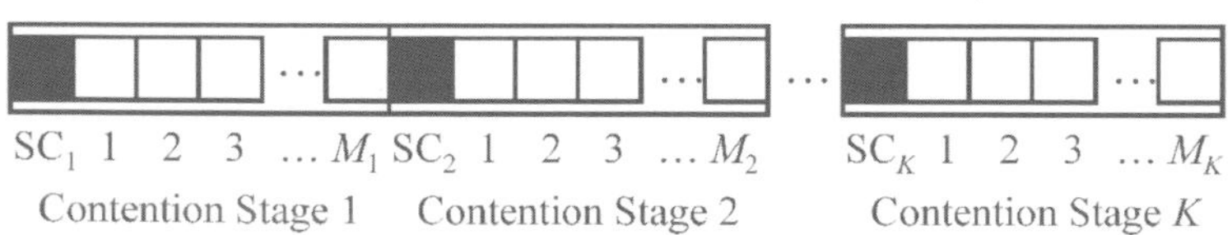


*Figure A.4.* Multi-stage contention.

$N_2$, where $N_i$ is the number of successful contentions at the $i$'th stage). The contention algorithm will be terminated upon collision-free transmission of all the reservations in contentions.

The expected number of unsuccessful nodes at the $k$'th contention stage, $U_k$, is

$$U_k = N\left(\frac{e-1}{e}\right)^k \tag{A.4}$$

The expected number of total contention stages, found by solving $U_k = 1$ is

$$K = \frac{\ln(N)}{1 - \ln(e-1)} \tag{A.5}$$

The upper panel in Figure A.5 shows the average number of contention stages obtained from simulation and theory. The expected number of total contention slots required for the termination of the algorithm, $S$, is

$$S = N\sum_{i=0}^{K-1}\left(\frac{e-1}{e}\right)^i \cong Ne \tag{A.6}$$

The lower panel in Figure A.5 shows the average values obtained from simulation and theory for the total number of contention slots required, $S$. The total time for "all successful" contention of $N$ nodes, $T$, is

$$T = NeT_S + KT_{SC} \tag{A.7}$$

The total number of contention slots required for the successful contention of 25 nodes is 64 with the multi-stage algorithm. Using the single-stage algorithm, only an average of 17 nodes can make successful contentions in 64 contention slots. The number of successful contentions reaches 24 with 600 contentions slots by using the single-stage contention algorithm. On average, 100% success is not possible with the single-stage algorithm, even with 1024 contention slots. However, 100 % success is realizable with the multi-stage algorithm with 64 contention slots and 7 contention stages on the average. The total contention duration for the multi-stage algorithm is $64T_S + 7T_{SC}$. If we assume the SC and contention packet sizes are equal, then the total contention duration is $71T_S$.

## A.5 Discussion

In our system we assumed there is no capture, but by changing the content of the SC messages, the system can easily adapt to capture (*i.e.*, instead of sending the number of successful contentions, the list of successful contentions can be sent, which completely eliminates the ambiguity that arises due to capture). This increases the SC packet size, but the multi-stage algorithm will still perform better than single-stage S-ALOHA, in terms of successful transmissions per contention slot.

It is possible to design a single-stage S-ALOHA system with $Ne$ contention slots and $Ne$ nodes, which results in $N$ successful transmissions, on the average. However, $N$ is the average of an ensemble including members significantly less than $N$, which results in underutilization of the data sub-frame. The multi-stage algorithm guarantees $N$ successful transmissions, but its length is a statistical quantity around its mean, $Ne$.

Although we can estimate the average number of contending nodes based on the statistics of the transmission history, we do not know which nodes are transmitting (*i.e.*, we only know the number of the nodes). Thus, it is not possible to assign data slots deterministically to those nodes, and we need a statistical scheme to assign the data slots through a contention algorithm.

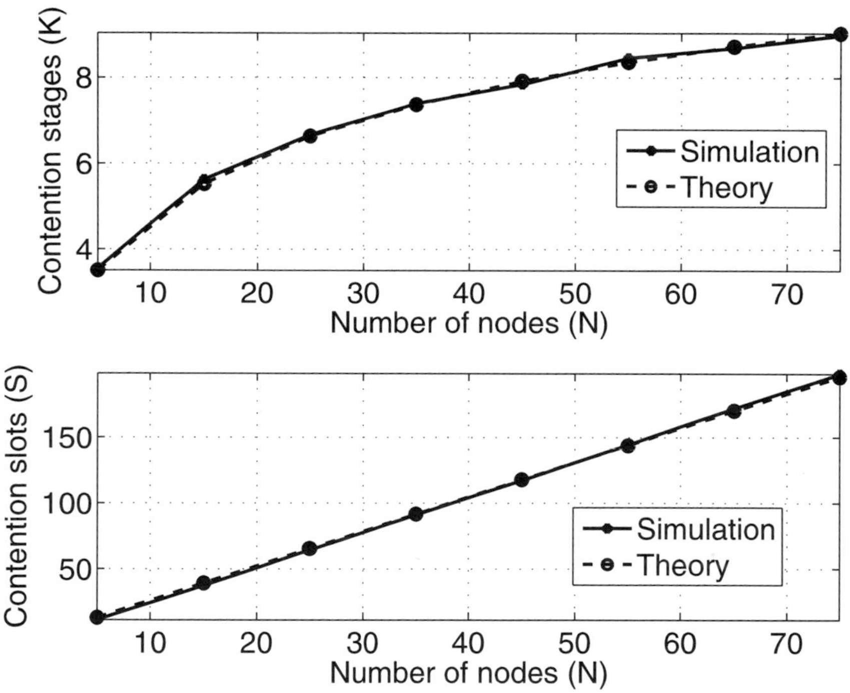


*Figure A.5.* The upper panel shows the total number of stages, K, as a function of the number of nodes, N. The lower panel shows the total number of contention slots required for the termination of the contention, S, as a function of N. Simulation results are the mean of 1000 independent runs.

## A.6 Summary

In this appendix we presented a multi-stage contention algorithm that results in the maximum number of successful contentions in minimum time for S-ALOHA type contention systems. Our analytical and simulation results show that our algorithm enables $N$ collision-free transmissions for $N$ nodes in $Ne$ contention slots on the average.

# Appendix B
# Effects of Clusterhead Separation on MH-TRACE

Minimum distance between neighboring clusterheads affects the MH-TRACE performance in terms of packet drops, collisions, and stability. However, the extent of the performance change as a function of minimum clusterhead separation is not so clear without actual measurements through simulations. In this appendix, we present the effects of minimum inter-clusterhead separation.

## B.1 Modified Cluster Creation and Maintenance Algorithms

We modified the cluster creation and maintenance algorithm, which are presented in Figure B.1 and Figure B.2, respectively, to investigate the effects of the minimum inter-clusterhead separation on protocol performance. In the modified cluster creation algorithm, if a node in startup mode does not hear any beacons but the interference level is higher than the maximum interference threshold, $Th_{IF}$, to start a new cluster, then the node is blocked from any transmissions, because it can neither become a clusterhead nor can it get channel access from a clusterhead due to the absence of clusterheads in its receive range. The maximum interference threshold is directly proportional with the distance (*i.e.*, the higher the $Th_{IF}$, the lower the minimum clusterhead separation). However, it can still receive all the packets in its receive range.

The rationale behind node blocking is that if a new cluster centered at the high interference region is created, then packet transmissions from the multiple clusters transmitting at the same time frame will collide at some locations with high probability. A blocked node always stays in the startup mode until the interference drops below the threshold or it starts to receive beacons from a clusterhead. To keep the consistency of the cluster creation algorithm, the cluster maintenance algorithm is also modified. A clusterhead in a high interference region resigns with a probability $p_{HI}$, which is set to 0.5. The reason

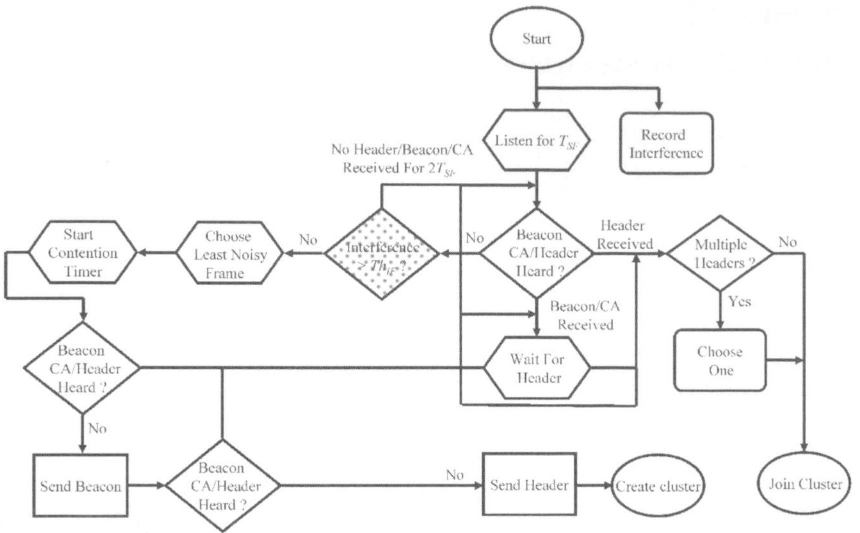


*Figure B.1.* MH-TRACE modified cluster creation algorithm flow chart. Modified blocks are marked with shaded background.

for adding a randomness in clusterhead resignation in such situations is that there are at least two clusterheads that create interference for one another. If both (or all) of them resign simultaneously, then there are more than enough resignations (*i.e.*, if there are two interfering clusterheads, resignation of only one of them is enough to resolve the interference problem), which will create inefficiency in the cluster maintenance algorithm.

## B.2 Simulation Results and Discussion

Table B.1 lists the maximum interference threshold, $Th_{IF}$ and corresponding minimum clusterhead separation distance, $D_{CH}$. We run simulations with 100 nodes moving within a 1 km by 1 km area for 100 s for different values of $N_F$ and $N_D$, which are listed in Table B.2. Note that in this appendix we set the carrier sense range to $\infty$ to observe the effects of inter-clusterhead separation without any constraints.

The average number of clusterheads at a time versus $D_{CH}$ is plotted in Figure B.3. The $x$-axis shows the minimum allowed co-frame clusterhead (clusterheads that use the same frame) separation distance ($D_{CH}$) and the $y$-axis shows the average number of clusterheads at a time. The curves in the figure are for each superframe configuration with a different number of superframes. 250 m $D_{CH}$ is the case where $Th_{IF}$ does not have any affect in the algorithm, because the minimum separation is actually equal to the transmit range. As expected, the average number of clusterheads, 10.5, is very close for $D_{CH} = 250$ m for

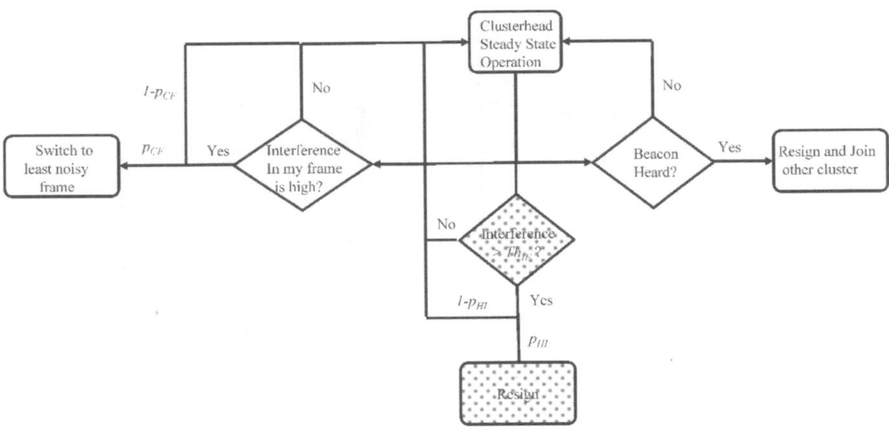


*Figure B.2.* MH-TRACE modified cluster maintenance algorithm flow chart. Modified blocks are marked with shaded background.

*Table B.1.* Minimum clusterhead separation and corresponding threshold.

| Minimum Clusterhead Separation (m) | Threshold (pW) |
|---|---|
| 250 | 365.2 |
| 350 | 95.1 |
| 450 | 34.8 |
| 550 | 15.6 |
| 650 | 8.0 |
| 750 | 4.5 |

*Table B.2.* Superframe parameters.

| Number of frames per superframe, $N_F$ | Number of data slots, $N_D$ | Number of contention slots, $N_C$ | Superframe time, $T_{SF}$, (ms) |
|---|---|---|---|
| 4 | 12 | 15 | 24.976 |
| 5 | 10 | 7 | 25.060 |
| 6 | 8 | 9 | 24.984 |
| 7 | 7 | 6 | 25.172 |
| 8 | 6 | 6 | 24.992 |

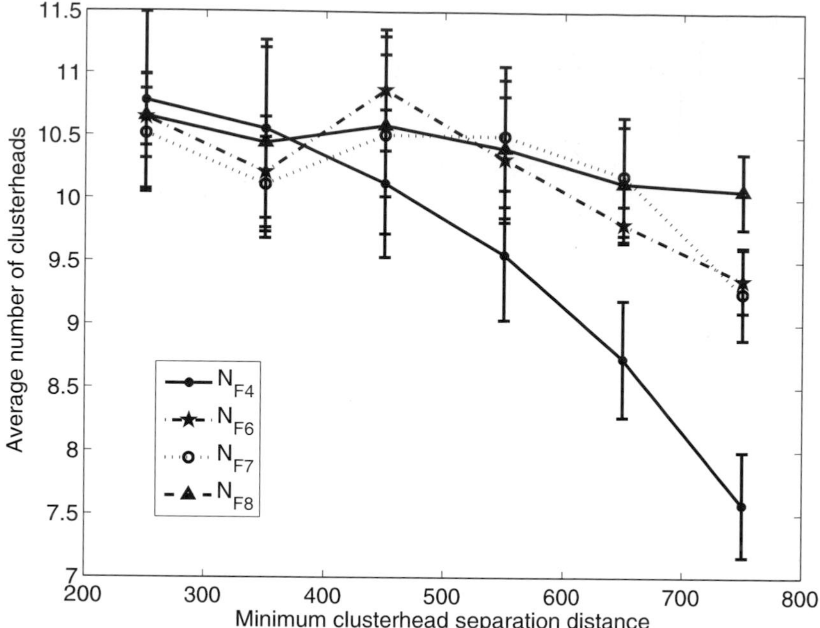


*Figure B.3.* Average number of clusterheads versus clusterhead separation.

all $N_F$, because the cluster creation algorithm becomes independent of $N_F$ for $D_{CH} = 250$ m. Uncertainties in the simulations due to the limited simulation time and finite ensemble set manifest themselves by the slight difference between the data points for $D_{CH} = 250$ m. The average number of clusterheads has a decreasing trend with increasing $D_{CH}$, because increasing $D_{CH}$ dictates more constraints in the clusterhead creation. The sharpest decrease is for $N_{F4}$ ($N_{F4}$ is used for $N_F = 4$) and the least decrease is for $N_{F6}$ and $N_{F7}$. $N_{F8}$ is almost not affected because for larger $N_F$ the clustering algorithm also becomes independent from $Th_{IF}$, because each node is surrounded by non-co-frame clusterheads and the minimum distance between the co-frame clusterheads automatically becomes large enough to avoid co-frame interference.

Figure B.4 shows the total number of clusterheads throughout the entire simulation time versus $D_{CH}$, which is an indicator of the stability of the clusters. The total number of clusterheads is not affected much from $N_F$ and $D_{CH}$, and it is in the vicinity of 30. An exception is $N_{F4}$ for $D_{CH} < 650$. In this range of $D_{CH}$, $N_{F4}$ creates twice the number of clusterheads that the other configurations create, and the standard deviation is very high when compared to other data points. Co-frame clusterheads should be well separated in space in order to avoid the situation that at some locations transmissions of co-frame clusterheads collide and no other clusterhead can be heard, which forces the nodes in these locations to enter startup and create their own clusters. Thus,

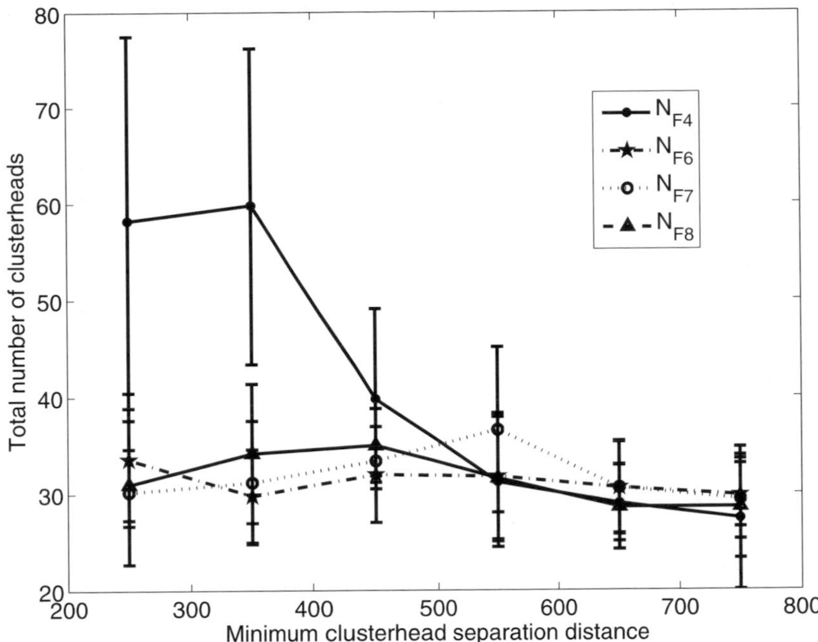


*Figure B.4.* Total number of clusterheads throughout the entire simulation time (100 s) versus clusterhead separation.

the clustering algorithm is not stable for $N_{F4}$ and $D_{CH} < 650$, and for $N_F > 4$ the stability of clustering algorithm is not affected from $D_{CH}$.

Figure B.5 shows the average number of blocked nodes versus $D_{CH}$. The number of blocked nodes is zero for $D_{CH} = 250$ m, because with this value of $D_{CH}$, the node blocking mechanism of the algorithm does not function. The number of blocked nodes increases with increasing $D_{CH}$. There are higher numbers of blocked nodes for lower frame numbers with increasing $D_{CH}$. $N_{F8}$ is the least affected by $D_{CH}$ and is also the least sensitive to $D_{CH}$ changes, as shown in Figure B.3.

Figure B.6 shows the average number of packets transmitted from the MAC layer per frame versus $D_{CH}$. The average number of packets generated per frame by the nodes is 42.9. With $D_{CH} = 250$ m, the average number of transmitted MAC packets for $N_{F5}$ and $N_{F8}$ is close to 41 and for $N_{F4}$, $N_{F6}$, and $N_{F7}$ is in the range $42.5 \pm 0.5$. The number of transmitted MAC packets decreases with increasing $D_{CH}$ due to the increasing number of blocked nodes (see Figure B.5). With $D_{CH} = 750$ m, the average number of transmitted MAC packets converges to 36.5 for $N_{F4}$ and $N_{F5}$, to 39.5 for $N_{F6}$, and to 40.5 for $N_{F7}$ and $N_{F8}$.

Figure B.7 shows the average number of collided packets per superframe versus $D_{CH}$. Observations from this figure are: (*i*) collisions decrease with

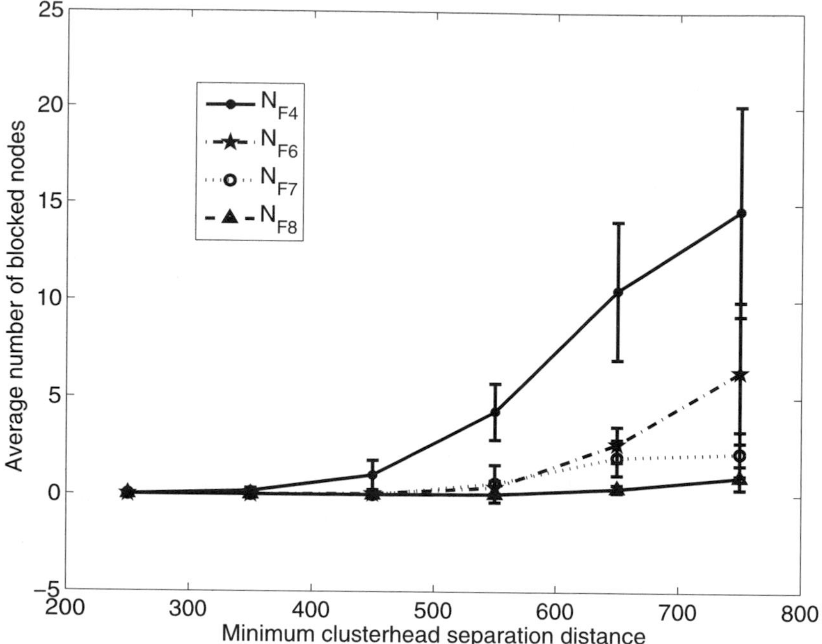


*Figure B.5.* Average number of blocked nodes per frame versus clusterhead separation.

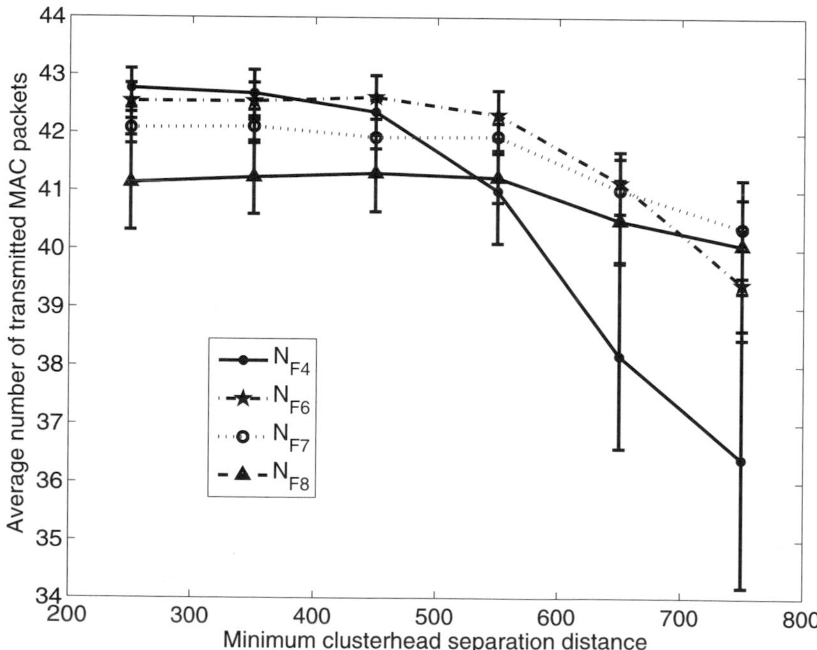


*Figure B.6.* Average number of transmitted MAC packets per superframe versus minimum clusterhead separation.

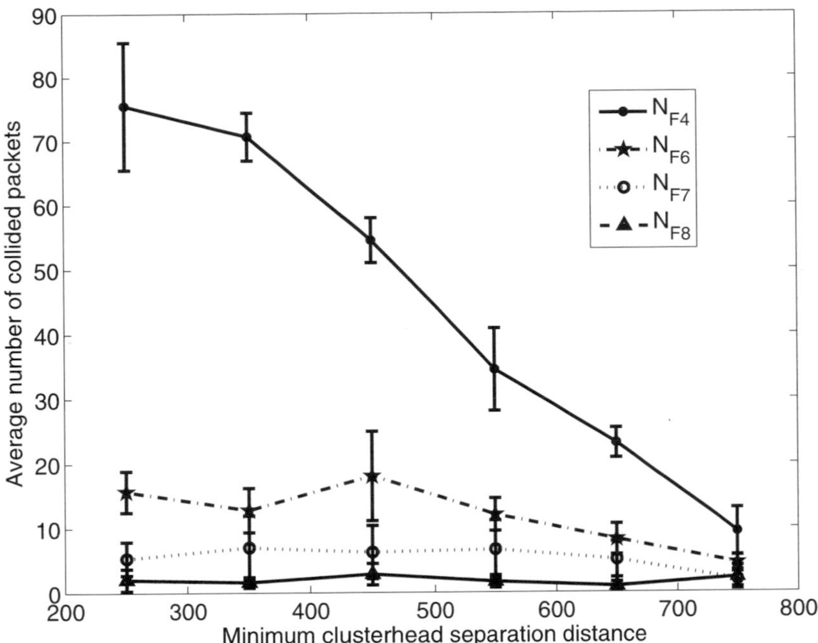


*Figure B.7.* Average number of collided packets per superframe versus minimum clusterhead separation.

increasing $D_{CH}$; (*ii*) higher frame number configurations have fewer collisions when compared to lower frame number ones; (*iii*) the number of collisions for $N_{F7}$ and $N_{F8}$ are almost insensitive to $D_{CH}$, and (*iv*) for $D_{CH} = 250$ m all the curves converge to $5 \pm 4$ interval. Collisions occur when the co-frame clusterheads are close. Thus, for $N_F < 6$ it is not possible to pull the number of collisions to a small marginal value without making use of $D_{CH}$.

Figure B.8 shows the average aggregate number of dropped packets per superframe versus minimum clusterhead separation. General trends in this figure are that (*i*) the number of dropped packets is higher for higher $D_{CH}$; (*ii*) for $D_{CH} = 250$ m, the number of dropped packets is pretty close for all $N_F$, and (*iii*) for higher $D_{CH}$ there are more packet drops for $N_{F4}$ and $N_{F5}$ and less for the others.

Figure B.9 shows the average aggregate number of received packets per superframe versus minimum clusterhead separation. Actually, this is the most important plot in this section, which shows the aggregate network throughput as a function of $N_F$ and $D_{CH}$. The bottom line is that the throughput is highest for $N_{F6}$ and $N_{F7}$ when $D_{CH} < 750$ m and it is low for $N_{F4}$ and $N_{F5}$. Increasing $N_F$ beyond seven does not increase the throughput but instead decreases it. Throughput is relatively insensitive to $D_{CH}$ for $D_{CH} < 650$ m, but it starts to decrease after this range.

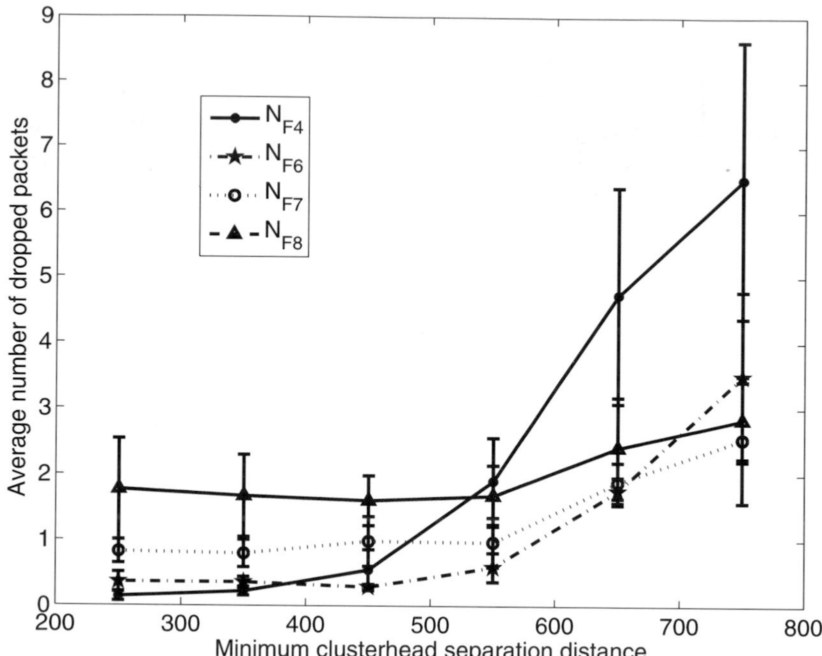


*Figure B.8.* Average number of dropped packets per superframe versus minimum clusterhead separation.

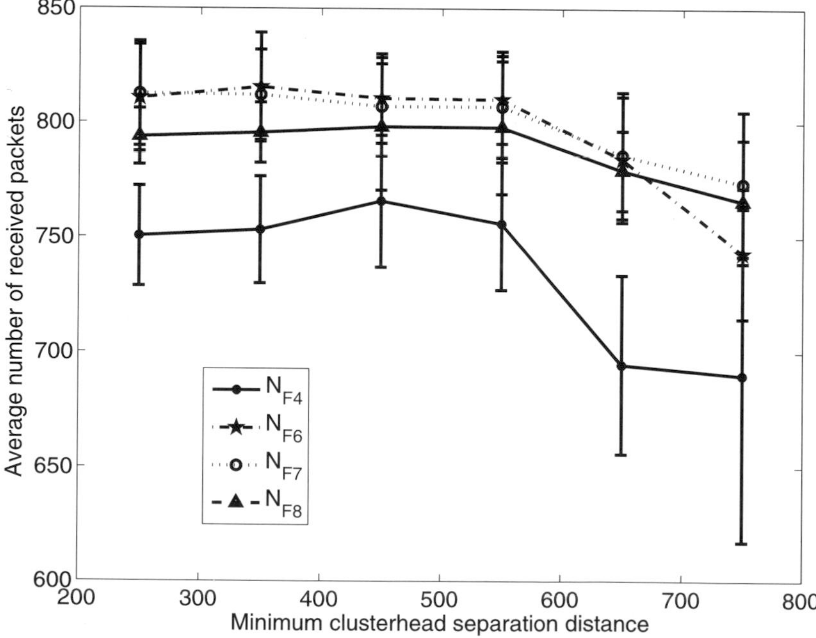


*Figure B.9.* Average aggregate number of received packets per superframe versus the minimum clusterhead separation.

## B.3 Summary

In summary, the conclusion we reach after analyzing the results of the simulations in this chapter is that the MH-TRACE cluster creation and maintenance algorithm presented in Chapter 7 is the best alternative. Thus, the interclusterhead separation does not need to be treated apart from the basic cluster creation and maintenance algorithm.

# Appendix C
# Broadcast Capacity of Wireless Ad Hoc Networks

In Chapter 9 we presented a comparative evaluation of MAC layers utilized in flooding. One of the important disadvantages of flooding is its spatial reuse inefficiency, thus, we designed the NB-TRACE protocol architecture (see Chapter 10) for better spatial reuse efficiency. In fact, the spatial reuse efficiency of NB-TRACE was shown to be far better than flooding. In other words, the broadcast capacity of NB-TRACE is higher than the broadcast capacity of flooding. Although the asymptotic bounds on the capacity of wireless ad hoc networks for unicasting are known, bounds on the broadcast capacity of wireless ad hoc networks are not known. Therefore, in this chapter, we present an upper bound on the broadcast capacity of arbitrary ad hoc wireless networks. We show that the throughput obtainable by each node for broadcasting to all of the other nodes in a network consisting of $n$ nodes with fixed transmission ranges and W bits per second channel capacity is bounded by $O(W/n)$, which is equivalent to the upper bound for per node capacity of a fully connected single-hop network.

## C.1 Background

The seminal work of Gupta and Kumar [180] has revealed that the per node capacity of ad hoc wireless networks decreases with increasing network size. They showed that the end-to-end per node capacity of an ad hoc network is $\Theta(1/\sqrt{n})$. In [181], it is shown that $\Theta(1/\sqrt{n}) = 0.047/\sqrt{n}$ for an ideally routed (*i.e.*, centralized control in the network layer), IEEE 802.11 MAC-based network. It was shown in [182] that by inserting access points connected by cables into an ad hoc network, per node capacity of the network could be kept constant (*i.e.*, $\Theta(1)$).

We will summarize the results of [180, 181]. Consider an ad hoc wireless network with channel capacity $W$ bits per second, area $A$ m$^2$, constant node density ($n_0$ nodes/m$^2$), and a total of $n$ nodes in the network, where each node has a fixed transmission radius $R$. Due to the spatial frequency reuse, the total one-hop bandwidth available in the network increases with network area. The upper bound on the gain from such spatial reuse is $O(A)$, which also can be expressed as $O(n)$ (*i.e.*, $n = An_0 \rightarrow O(A) = O(An_0) = O(n)$).

The average distance between randomly chosen source and destination pairs is proportional to the square root of the network area, which can also be expressed as the square root of $n$ (*i.e.*, $O(\sqrt{n})$). Thus, on the average, each bit should be relayed by $O(\sqrt{n})$ hops to its destination by the intermediate nodes on the path between the source and destination. This means that the aggregate bandwidth required to transfer each generated bit from the source to the destination is $O(\sqrt{n})$ bits per second.

If we model the multi-hop network as a fully connected single hop network, then due to spatial reuse the aggregate network bandwidth is increased by $O(n)$; and due to multi-hop relaying the bandwidth required to send a bit from source to the destination is increased by $O(\sqrt{n})$. When we combine these two mechanisms, the single-hop equivalent aggregate bandwidth of a multi-hop network as a function of $n$, $W^{ag}_{mh}(n)$, is obtained as

$$W^{ag}_{mh}(n) = \overbrace{O(n)}^{spatial\ reuse} \times \overbrace{O(1/\sqrt{n})}^{multi-hop\ relaying} \times \overbrace{W_{sh}}^{channel\ capacity} \qquad \text{(C.1)}$$

where $W_{sh}$ is the channel capacity, $W$ bits per second. This aggregate capacity is characterized as:

$$W^{ag}_{mh}(n) = O(W\sqrt{n}) \qquad \text{(C.2)}$$

The per node capacity of the network, $W^{pn}_{mh}(n)$is:

$$W^{pn}_{mh}(n) = W^{ag}_{mh}(n)/n = O(W/\sqrt{n}) \qquad \text{(C.3)}$$

The theoretical limits on the capacity of ad hoc wireless networks discussed so far are for unicast traffic (*i.e.*, one-to-one). The broadcast capacity of arbitrary ad hoc wireless networks has not been investigated in the literature. The main reason for the lack of attention to this problem is that multi-hop broadcasting is not the main service targeted in ad hoc networks. However, in some ad hoc and sensor network applications, network-wide broadcasting is the primary function of the network. Furthermore, all the routing protocols for unicasting use broadcasting for route discovery, monitoring, and maintenance. Thus, the limitations imposed by broadcasting are crucial in the analysis of unicast routing protocol architectures used in ad hoc and sensor networks as well.

## C.2 Upper Bound on Broadcast Capacity

In unicasting, the average path length of randomly chosen source-destination pairs is related with the square root of the network area, $\sqrt{A}$. However, in broadcasting all the nodes in the network should receive each packet. Thus, the path length in broadcasting is related with the network area, $A$, instead of $\sqrt{A}$ in unicasting, whereas the spatial reuse factor in broadcasting is the same as in unicasting.

An upper bound on the single-hop equivalent aggregate bandwidth of a multi-hop network in broadcasting as a function of $n$, ${}^{bc}W_{mh}^{ag}(n)$, is given as

$$ {}^{bc}W_{mh}^{ag}(n) = \overbrace{O(n)}^{spatial\ reuse} \times \overbrace{O(1/n)}^{multi-hop\ relaying} \times \overbrace{W_{sh}}^{channel\ capacity} \tag{C.4} $$

Note that the multi-hop relaying term for broadcasting is $O(1/n)$, whereas in unicasting it was $O(1/\sqrt{n})$. Thus, the aggregate throughput capacity for broadcasting in a multi-hop network is bounded by

$$ {}^{bc}W_{mh}^{ag}(n) = O(W) \tag{C.5} $$

Per node capacity for broadcasting is bounded by

$$ {}^{bc}W_{mh}^{pn}(n) = {}^{bc}W_{mh}^{ag}(n)/n = O(W/n) \tag{C.6} $$

To support the above intuitive analysis of broadcast capacity, we will formally establish an upper bound on the broadcast capacity of arbitrary ad hoc networks.

**Theorem 1**: The upper bound on the per node broadcast capacity of an arbitrary ad hoc network is $O(1/n)$.

We will provide two alternative proofs for theorem 1.

**Proof 1-1**: Assuming a constant transmit radius, $r_0$, for each node in the network, the coverage area of each node, $A_0$, is $\pi r_0^2$. Thus, any transmission can be received by at most $A_0 n_0$ number of nodes. To cover the entire network, which is the goal in network-wide broadcasting, at least $A/A_0$ transmissions are required.

As an extreme case, assume perfect capture, where a receiving node receives the higher power packet if there are multiple simultaneous packet transmissions by multiple transmitters. Therefore, any two transmitters must be separated by at least $2r_0$ to ensure that all the nodes in the receive range of each transmitter are receiving the packets destined for them. By considering the fact that a transmitting node can be in the corner, the maximum number of concurrent transmissions is then equal to $A/(A_0/4)$.

When we combine these two results we see that each generated bit needs to be retransmitted at least for $[A/A_0 - 1]$ times, and it is possible to transmit at most $A/(\pi r_0^2/4)$ bits concurrently. Therefore, the aggregate broadcast capacity that can be supported is:

$$W[A/(A_0/4)]/[A/A_0] = W(4A_0/A_0) = 4W \qquad \text{(C.7)}$$

Per node broadcast capacity is obtained as $4W/n = O(1/n)$ □.

**Proof 1-2**: Let the set $\mathrm{S_{MCDS}}$ denote the subset of nodes that create a Minimally Connected Dominating Set (MCDS) for the network. An MCDS is a minimal set of connected nodes such that any non-set node is in the one-hop neighborhood of at least one member of the set. An MCDS creates an optimal broadcasting (retransmission) scheme [20, 87]. Let the number of nodes in an MCDS be $n_1$. Since each node in $\mathrm{S_{MCDS}}$ has to transmit at least once, the total number of transmissions required for a packet to be broadcast to the entire network is $n_1$ for any source node within the set, and the number of transmissions is $n_1+1$ for any non-set node. The maximum number of simultaneous successful transmissions within the MCDS is $n_1/2$, because each downstream node should be listening to the upstream node to keep the broadcast flow alive. Thus, the aggregate bandwidth is bounded by

$$W(n_1/2)/(n_1+1) \overset{\lim_{n\to\infty}}{=} W/2 \qquad \text{(C.8)}$$

The per node broadcast capacity is obtained as $W/2n = O(1/n)$, which concludes the proof □.

## C.3 Summary

We present an upper bound on the broadcast capacity of arbitrary ad hoc wireless networks. The throughput obtainable by each node for broadcasting to all of the other nodes in a network consisting of $n$ nodes with fixed transmission ranges and $W$ bits per second channel capacity is bounded by $O(W/n)$, which is equivalent to the upper bound for per node capacity of a fully connected single-hop network. Thus, the scalability of broadcasting is worse than unicasting and the scalability of multicasting is in between them. Depending on the multicast group size, per node broadcast capacity of multicasting can be either $O(W/n)$, if the multicast group size is not bounded, or $O(W/\sqrt{n})$, if the multicast group size is bounded by a finite number.

# Appendix D
# Glossary of Terms

| | |
|---|---|
| **ABR** | Available Bit Rate |
| **ACB** | Activate Branch |
| **ACK** | Acknowledgment |
| **AMRIS** | A Multicast pRotocol for ad hoc wIreleSs networks |
| **ARN** | Average number of Retransmitting Nodes per data packet |
| **ASK** | Amplitude Shift Keying |
| **BEB** | Binary Exponential Backoff |
| **BER** | Bit Error Rate |
| **BFSK** | Binary Frequency Shift Keying |
| **BPSK** | Binary Phase Shift Keying |
| **BRAN** | Broadband Radio Access Network |
| **BTMA** | Busy Tone Multiple Access |
| **CBB** | Counter-Based Broadcasting |
| **CBR** | Constant Bit Rate |
| **CDMA** | Code Division Multiple Access |
| **CDS** | Connected Dominating Set |
| **CRB** | Create Branch |
| **CRC** | Cyclic Redundancy Check |
| **CSMA** | Carrier Sense Multiple Access |
| **CTS** | Clear To Send |
| **DBB** | Distance-Based Broadcasting |
| **DCF** | Distributed Coordination Function |

| | |
|---|---|
| **DECT** | Digital European Cordless Telephone |
| **DIFS** | Distributed Inter-Frame Space |
| **DMAC** | Distributed and Mobility Adaptive Clustering |
| **DR-TDMA** | Dynamic Reservation TDMA |
| **DSDV** | Destination-Sequenced Distance Vector routing |
| **DSSS** | Direct Sequence Spread Spectrum |
| **DSR** | Dynamic Source Routing |
| **EC-MAC** | Energy Conserving Medium Access Control |
| **FDMA** | Frequency Division Multiple Access |
| **FEC** | Forward Error Correction |
| **FHSS** | Frequency Hopping Spread Spectrum |
| **FSK** | Frequency Shift Keying |
| **FSM** | Frame Synchronization Message |
| **GPS** | Global Positioning System |
| **GSM** | Global System for Mobile communication |
| **HDTS** | Hop Distance To Source |
| **HRPGM** | Hierarchical Reference Point Group Mobility model |
| **HSDPA** | High Speed Downlink Packet Access |
| **IFL** | Initial Flooding |
| **IFS** | Inter-Frame Space |
| **IP** | Internet protocol |
| **ITU** | International Telecommunications Union |
| **JIT** | Just-In-Time |
| **LCC** | Least Cluster Change |
| **LNA** | Low Noise Amplifier |
| **MAC** | Medium Access Control |
| **MACA** | Multiple Access Collision Avoidance |
| **MACAW** | Multiple Access Collision Avoidance for Wireless |
| **MCDS** | Minimally Connected Dominating Set |
| **MC-TRACE** | Multicasting through TRACE |
| **MFSK** | Multiple Frequency Shift Keying |
| **MH-TRACE** | Multi-Hop TRACE |
| **MH-TRACE-NES** | MH-TRACE with No Energy Savings |
| **MNB** | Maintain Branch |
| **MPEG** | Motion Picture Expert Group |
| **MTS** | Multicast Tree Size |

| | |
|---|---|
| **ns** | Network Simulator |
| **NAV** | Network Allocation Vector |
| **NB-TRACE** | Network-wide Broadcasting through TRACE |
| **NCDS** | Non-Connected Dominating Set |
| **NIC** | Network Interface Card |
| **NTDR** | Near Term Digital Radio |
| **ODMRP** | On Demand Multicast Routing Protocol |
| **OSI** | Open System Interconnection |
| **PCB** | Partially Coordinated Broadcast algorithms |
| **PDA** | Personal Digital Assistant |
| **PCM** | Pulse Code Modulation |
| **PDR** | Packet Delivery Ratio |
| **PRN** | Pruning |
| **PRMA** | Packet Reservation Multiple Access |
| **PDF** | Probability Density Function |
| **PSK** | Phase Shift Keying |
| **RPB** | Repair Branch |
| **QoS** | Quality of Service |
| **R-ALOHA** | Reservation-ALOHA |
| **RAD** | Random Assessment Delay |
| **RCPC** | Rate Compatible Punctured Convolutional encoding |
| **RMS** | Root Mean Square |
| **RPGM** | Reference Point Group Mobility |
| **RTP** | Real-Time Protocol |
| **RTCP** | Real-Time transport Control Protocol |
| **RTS** | Request To Send |
| **RWP** | Random Way Point mobility model |
| **S-ALOHA** | Slotted ALOHA |
| **SMAC** | Sensor-MAC |
| **SDMA** | Space Division Multiple Access |
| **SIFS** | Short Inter-Frame Space |
| **SH-TRACE** | Single-Hop TRACE |
| **SNR** | Signal-to-Noise Ratio |
| **TCP** | Transmission Control Protocol |
| **TDMA** | Time Division Multiple Access |
| **TRACE** | Time Reservation using Adaptive Control for Energy efficiency |

**UDP** User Datagram Protocol

**VBR** Variable Bit Rate

**VAD** Voice Activity Detector

**WHD** Weighted Highest Degree clustering

# Index

# About the Authors

**Bulent Tavli** received the B.S. and M.S. degrees in electrical and electronics engineering in 1996 from Middle East Technical University, Ankara, Turkey and in 1998 from Baskent University, Ankara, Turkey. He received the M.S. and Ph.D. degrees in electrical and computer engineering in 2001 and 2005 from the University of Rochester, Rochester, NY. Currently he is a post-doctoral researcher at the University of Rochester. His research interests lie in the areas of wireless communications and networking, ad hoc and sensor networks, signal and image processing, and biomedical ultrasound. Dr. Tavli is a member of the IEEE.

**Wendi Heinzelman** is an assistant professor in the Department of Electrical and Computer Engineering at the University of Rochester. She received a B.S. degree in Electrical Engineering from Cornell University in 1995 and M.S. and Ph.D. degrees in Electrical Engineering and Computer Science from MIT in 1997 and 2000, respectively. Her current research interests lie in the areas of wireless communications and networking, mobile computing, and multimedia communication. Dr. Heinzelman received the NSF CAREER award in 2005 for her work on cross-layer architectures for wireless sensor networks, and she received the ONR Young Investigator Award in 2005 for her work on balancing resource utilization in wireless sensor networks. She is a member of Sigma Xi, the IEEE, and the ACM.